Sounding Bodies

SUNY series, Studies in the Long Nineteenth Century

Pamela K. Gilbert, editor

Sounding Bodies

Acoustical Science and Musical Erotics
in Victorian Literature

SHANNON DRAUCKER

SUNY
PRESS

Cover: Thomas Wilmer Dewing, "A Musician," 1877.

Published by State University of New York Press, Albany

© 2024 State University of New York

All rights reserved

Printed in the United States of America

No part of this book may be used or reproduced in any manner whatsoever without written permission. No part of this book may be stored in a retrieval system or transmitted in any form or by any means including electronic, electrostatic, magnetic tape, mechanical, photocopying, recording, or otherwise without the prior permission in writing of the publisher.

Links to third-party websites are provided as a convenience and for informational purposes only. They do not constitute an endorsement or an approval of any of the products, services, or opinions of the organization, companies, or individuals. SUNY Press bears no responsibility for the accuracy, legality, or content of a URL, the external website, or for that of subsequent websites.

For information, contact State University of New York Press, Albany, NY
www.sunypress.edu

Library of Congress Cataloging-in-Publication Data

Name: Draucker, Shannon, 1991– author.
Title: Sounding bodies : acoustical science and musical erotics in Victorian literature / Shannon Draucker.
Description: Albany : State University of New York Press, [2024]. | Series: SUNY series, studies in the long nineteenth century | Includes bibliographical references and index.
Identifiers: LCCN 2023050595 | ISBN 9781438498416 (hardcover : alk. paper) | ISBN 9781438498393 (ebook)
Subjects: LCSH: English literature—19th century—History and criticism. | Music in literature. | Eroticism in literature. | Human body in literature. | Music—Physiological effect. | Feminist theory. | Queer theory. | LCGFT: Literary criticism.
Classification: LCC PR468.M857 D73 2024 | DDC 820.9/353809034—dc23/eng/20240229
LC record available at https://lccn.loc.gov/2023050595

For Claire and Carl Draucker

Common experience teaches us that all sounding bodies are in a state of vibration. This vibration can be seen and felt; and in the case of loud sounds we feel the trembling of the air even without touching the sounding bodies.

—Hermann von Helmholtz,
Popular Lectures on Scientific Subjects (1881)

Contents

List of Illustrations — ix

Acknowledgments — xi

Introduction   The Erotic Symphony — 1

Part One: Sounds and Bodies

Chapter One   Hearing, Touching, Feeling Sound: Acoustical Science in Nineteenth-Century Britain — 25

Part Two: Genders

Chapter Two   Bare Arms and Quivering Nerves: The "Lady Violinist" Novels of Mary Augusta Ward and M. E. Francis — 51

Chapter Three   Cross-Dressing Violinists and Music/Gender Performance in *The Heavenly Twins* and *The Violin-Player* — 79

Part Three: Sexualities

Chapter Four   Dangerous Vibrations: Musical Rape in George Eliot and Thomas Hardy — 101

| Chapter Five | Orgasm in the Orchestra Box: *Teleny*'s Musical Pornography | 127 |

Part Four: Intimacies

Chapter Six	Fiddle Feelings: Human-Instrument Intimacies in Dickens, Eliot, Trollope, and Hardy	147
Chapter Seven	Musical Hauntings and Otherworldly Erotics in *The Lost Stradivarius* and "A Wicked Voice"	169
Coda	Re-vitalizing Contemporary Classical Music	189
Notes		199
Bibliography		245
Index		271

Illustrations

Figure 1.1	Valentine for John Tyndall, "Sonorous Sensitive!!!!"	26
Figure 1.2	Illustration of Chladni plates, 1867.	28
Figure 1.3	Tyndall's glass ball demonstration, 1867.	30
Figure 1.4	Tyndall's human sound propagation diagram, 1867.	30
Figure 1.5	"Acoustic Instruments," 1819.	33
Figure 1.6	Curtis's acoustic chair, 1842.	33
Figure 1.7	Helmholtz resonator, 1863.	35
Figure 1.8	Exercises, in Flora T. Parsons, *Callisthenic Songs Illustrated*, 1869.	46
Figure 2.1	Painting by James Tissot, *Hush!*, c. 1874.	55
Figure 2.2	Cartoon by George Du Maurier, "The Fair Sex-Tett," c. 1875.	58
Figure 2.3	Painting by Thomas Wilmer Dewing, *Lady with Cello*, 1920 or before.	58
Figure 2.4	Cartoon, "Joachim's Rival," c. 1880.	59
Figure 2.5	Exercise, in Anna Leffler Arnim, *A Complete Course of Wrist and Finger Gymnastics*, 1894.	68
Figure 5.1	"Viola da Gamba" by Franz von Bayros, c. 1908.	131
Figure 6.1	Advertisement for the technicon, c. 1887.	154

Acknowledgments

At its core, *Sounding Bodies* is about the profound pleasures of artistic and intellectual exchanges. Many such exchanges have sustained my work on this book, and I am endlessly grateful.

I first began to think about music and literature in tandem while writing my undergraduate honors thesis at Dartmouth College, where Andrew McCann introduced me to *The Lost Stradivarius*, *Teleny*, and other tantalizingly strange nineteenth-century texts and taught me how to conduct scholarly research. Taking classes with Melanie Benson Taylor and Aimee Bahng made me want to be a professor, and their sound and supportive guidance as I applied to graduate school made me feel like it might be possible.

As a PhD student at Boston University, my intellectual world opened up in ways I never could have imagined. I owe my deepest thanks to Julia Prewitt Brown, whose enthusiasm for this project from our earliest conversations about Dickens and music gave me the confidence to find my scholarly voice. I am fortunate to be one of the many students whose lives have been enriched by Julia's brilliance, warmth, and generosity. Anna Henchman not only sparked my unexpected interest in the history of science but also urged me to make bold claims and reminded me that imagination and wonder are crucial components of academic work. As both a dissertation reader and teaching mentor, Carrie Preston shaped how I think about gender, sexuality, and pedagogy. She validated my queer and reparative readings and modeled a radically inclusive, feminist approach to teaching that I seek to emulate in my own classrooms. I had the great luck to briefly overlap with Susan Bernstein during my final year at BU, and her incisive feedback improved my work immensely. I am especially grateful to Joe Rezek for helping me navigate this project's bumpy transition from dissertation to book—and my own (at times also bumpy) transition from graduate student to professor.

At BU, I found myself surrounded by some of the smartest, strongest, and most fun women I have ever met, including Chelsea Bray, Agnes Burt, Pardis Dabashi, Reed Gochberg, AJ Gold, Emily Gowen, Annael Jonas-Paneth, and Talia Vestri. Thank you for the shared meals, study dates, work exchanges, and gossip sessions. An extra thanks to Reed for her help with so many iterations of this project, from dissertation abstract to book proposal.

Though I often read and write alone, I am rarely lonely in my academic work thanks to a wonderful community of scholars around the world. My completion of this book owes a great deal to Kate Nesbit, Amanda Shubert, Doreen Thierauf, and Kim Cox. Kate and Amanda have read nearly every word of this manuscript and are among my most cherished co-thinkers. Amanda created the writing group that introduced me to Grace An, Ally Field, Ari Gass, and Katerina Korola; thank you all for the weekly camaraderie, accountability, and conversations about pets and food—and for propping me up on some especially tearful days. Doreen is the world's sharpest editor and thinker—her feedback has improved nearly every chapter of this book—and one of my dearest fellow wanderers through "early career" life. Kim helped me reframe the introduction at a crucial stage, and our collaboration, together with Doreen, on a special issue of *Nineteenth-Century Gender Studies* helped me survive the long pandemic winter. Recently, Doreen, Kim, Kate, and I joined forces with Riya Das and Ashley Nadeau to advocate for greater attention to teaching in the field of Victorian studies. I am energized by our feminist solidarity and shared labor. My gratitude extends to countless other friends and mentors in the field, among them Nathan Hensley, whose feedback on the book proposal and reassurance about the project pulled me out of a rut at a moment when I felt like giving up. Thanks also to the members of the Nineteenth-Century Seminar working group at Dickens Universe in 2016 for creating a safe space for a nervous graduate student to start exploring these ideas.

Several grants and fellowships facilitated my work on *Sounding Bodies*. A National Endowment for the Humanities Summer Stipend in 2022 gave me the time essential to complete the book. At Siena College, the Fr. Peter Fiore, O.F.M. Excellence in English Endowment and the Diversity Action Committee Diversity Research Fellowship provided crucial support. Two Boston University Graduate Research Abroad Fellowships enabled generative archive trips to the UK. The Midwest Victorian Studies Association's Walter L. Arnstein Prize (Honorable Mention) also assisted with dissertation research. A Huntington Library short-term fellowship gave me four heavenly weeks to read and think among the cacti. Special thanks to the archivists at the

Huntington, the Royal College of Music, the Royal Academy of Music, the British Library, the Royal Institution, the Thomas Hardy Archives, the Wellcome Library, the Women's Library, and the Kinsey Institute for being such welcoming and enthusiastic interlocutors.

I am thrilled that this book found a home at SUNY Press, where I have benefited tremendously from Rebecca Colesworthy's expert guidance. Many thanks to Pamela Gilbert for including *Sounding Bodies* in SUNY's Studies in the Long Nineteenth Century series. I am also greatly appreciative of the two anonymous peer reviewers for their exceedingly helpful feedback on this manuscript. Portions of the acoustical research presented in chapter 1 appeared in "Hearing, Sensing, Feeling Sound: On Music and Physiology in Victorian England," *BRANCH: Britain, Representation, and Nineteenth-Century History* (June 2018), as well as in " 'Vibrating Through All Its Breadth': Musical Fiction and Materialist Aesthetics in the *Strand Musical Magazine*," *Victorian Periodicals Review* 51, no. 1 (Spring 2018): 1–17, © The Research Society for Victorian Periodicals. An earlier version of chapter 2 was published in "Performing Power: Female Musicianship and Embodied Artistry in Bertha Thomas's *The Violin-Player*," *Nineteenth-Century Gender Studies* 14, no. 1 (Spring 2018). Chapter 5 is derived in part from the article "Music Physiology, Erotic Encounters, and Queer Reading Practices in *Teleny*," *Victorian Literature and Culture* 50, no. 1 (Spring 2022): 141–72, reproduced with permission. Portions of the research on female performance and concert etiquette that appears in chapter 1, chapter 2, and the coda were published in "Ladies' Orchestras and Music-as-Performance in *fin-de-siècle* Britain," *Nineteenth-Century Contexts* 45, no. 1 (2023): 7–22, © Taylor & Francis, available online: https://doi.org/10.1080/08905495.2023.2161845. A piece I wrote for the *LA Review of Books* blog in summer 2021, titled "Food for the Soul, Art of the Flesh: Classical Music, COVID-19, and the Body," gave me the opportunity to think through some of the broader ideas in my coda. My sincere thanks to all of the above publications for the permission to reprint material.

My colleagues at Siena College have made Loudonville, New York, a lovely place to land. Their collegiality, generosity, and good humor make work a joy. Thanks to the staff at the Standish Library, and especially Sarah Symans, for their patience with my many Interlibrary Loan requests (and even a contactless book drop during a COVID-19 infection!). To my students: your insights, energy, love of literature, and passion for social justice invigorate me every single day. Lindsay Perrillo assisted with the final stages of research, and I am grateful to Siena's Center for Undergraduate

Research and Creative Activity (CURCA) for supporting her work through the Summer Scholars program. Thanks to the students in Sexuality in Literature and Women in Literature for always pushing me to think about gender and sexuality from new angles and to the students in my Narratives of Sexual Violence seminars in 2022 and 2023 for the conversations that helped shape chapter 4.

My friends and family members deserve my endless gratitude for their patience and encouragement throughout this process—and for listening to countless monologues about nineteenth-century ear diagrams and women violinists. Thanks to my Albany community for helping me build a life in a new city during a global pandemic. Danielle Dorfman has heard me prattle on about books and music for nearly twenty-five years; our friendship is the greatest gift. Though I do not have siblings, Lindsey Romero has been my sister for many years, and I cherish her wisdom, tough love, and boundless capacity for care. Anne Burke will always be my favorite fellow bookworm; thank you for reading every single piece I ever publish and for introducing me to all of the good cozy mysteries.

"How could the musical people avoid falling in love with each other?" George Eliot writes in *The Mill on the Floss*. Matt Boyas and I fell in love in our college orchestra's clarinet section nearly fifteen years ago, and his partnership has sustained me ever since. Matt enables my writing in countless ways both tangible (coffee, meals, housework, tech help) and intangible (patience when I go into my "writing holes," gentle nudges to crawl out of them when needed). Hazel the labradoodle came bounding into our lives in early 2020, and she has proven a source of both delightful distraction and cherished companionship ever since.

This book is dedicated to my parents, Claire and Carl Draucker, who have always encouraged me to read voraciously, think deeply, and dream wildly. They have filled my life with literature and music (and sat through more long orchestra concerts than anyone should ever have to!). My mom is also a professor, and my lifelong admiration of her work meant that my pursuit of an academic career was probably inevitable. From letting me run around the Kent State University campus as a kid, to teaching me about the virtues of a "coffee edit" (a term that, to my delight, some of my own students now use), to modeling what a career devoted to rigorous scholarship and compassionate mentorship looks like, my mom exposed me to the joys and challenges of an academic life and gave me the tools to navigate them. I owe her everything.

Introduction

The Erotic Symphony

> I, with the rest,
> Sat there athirst, atremble for the sound;
> And as my aimless glances wandered round,
> Far off, across the hush'd, expectant throng,
> I saw your face that fac'd mine.
> Clear and strong
> Rush'd forth the sound, a mighty mountain stream;
> Across the clust'ring heads mine eyes did seem
> By subtle forces drawn, your eyes to meet.
> Then you, the melody, the summer heat,
> Mingled in all my blood and made it wine.
> Straight I forgot the world's great woe and mine . . .
>
> —Amy Levy, from "Sinfonia Eroica: To Sylvia" (1884)

Is the concert hall more erotic than the bedroom? Amy Levy's "Sinfonia Eroica" seems to suggest so. The thirty-nine-line poem unfolds as the speaker attends a performance of Beethoven's Third Symphony on a "drowsy, golden afternoon" in June.[1] As the music swells and swirls throughout the hall, the speaker locks eyes with the addressee ("Sylvia") and admires her "body fair" and "perfect throat."[2] Yet Levy's speaker is just as aroused by the music itself as by the sight of their lover's body. They await the first notes "athirst, atremble for the sound" and "quiver[]" as the symphonic strains rise and fall.[3] It is Sylvia *and* "the melody" *and* "the summer heat" that "mingle" in the speaker's "blood."[4]

Scholars often read "Sinfonia Eroica" as a representation of same-sex desire (most gender the speaker as female) that can be experienced only through a longing glance across a crowded room. Critics not only point to the tantalizing wordplay offered by the subtitle of Beethoven's symphony ("Eroica") but also highlight music's figurative role in the poem, as an "extended metaphor for love-making," a symbol of "implied sexual climax," a vehicle to "safe[ly]" express "passions in public," or, more somberly, a representation of queer sadness in which the music's ephemerality ties same-sex desire to "absence," "negation," "doubt," and "delusion."[5]

However, "Sinfonia Eroica" also invites less figurative—and more reparative—readings, to use Eve Kosofsky Sedgwick's term for analyses that center "the many ways in which selves and communities succeed in extracting sustenance from the objects of a culture, even of a culture whose avowed desire has often been not to sustain them."[6] Levy's poem presents not simply a euphemism for sexual contact desired but denied, but also a concrete evocation of the physical and physiological experience of listening to music—an experience that is itself erotic and shared among the entire "thronging" crowd in the concert hall.[7] Music listening is a luscious sensory event in its own right; Beethoven's symphony stirs the blood and ignites the pulses, making the very air quiver with delight and bringing bodies—both human and nonhuman—into pleasurable relation.

The context of nineteenth-century acoustical science, flourishing when Levy was writing her poem, urges such a concretely physiological interpretation. Amid the mid-nineteenth-century "English Musical Renaissance," which witnessed a flood of symphony orchestras, concert halls, and conservatories in cities across Britain, scientists like the German physician and physicist Hermann von Helmholtz and the British physicist John Tyndall researched how and why this newly omnipresent music so deeply affected its listeners and players.[8] While eighteenth-century and Romantic philosophers often conceptualized music as an ineffable, transcendent entity, Victorian acousticians mobilized the tools of experimental science to reveal music's material properties. Embraced by thinkers across a range of scientific fields—including physics, physiology, anatomy, biology, evolutionary science, psychology, and medicine—nineteenth-century acoustical science centered on two main areas of study: "physical acoustics," which focused on the material processes of sound transmission, and "physiological acoustics," which explored the effects of sound on the ear and the human body.[9] Sound, acousticians discovered, is a physical entity composed of particle-filled waves that vibrate the air, tickle the nerves of the ear, and activate the body's muscular and circulatory systems.[10] They argued that music in particular arose from waves that vibrated especially regularly and frequently.[11] It is due to nineteenth-century acoustics that scientists now understand, for instance, why an opera singer's voice can break glass, why dancers feel propelled to move when the beat drops in a nightclub, and why ASMR videos "tingle" some viewers' brains and cause goosebumps.

By the late nineteenth century, as Helmholtz biographer Benjamin Steege writes, there was a "robust . . . market for popular acoustics in a Helmholtzian vein," particularly in Victorian England.[12] Articles about acoustics appeared in periodicals ranging from *Nature* to *Punch*, and scientists gave wildly popular public lectures on the subject at venues like the Royal Institution and the Athenaeum and at universities across Britain.[13] Once thought of as an abstract, intangible phenomenon that transcended the physical world and sent the spirit soaring, music came to be understood as a physical entity that could be studied and quantified and that could affect bodies and things in measurable ways.

These emerging discoveries in acoustical science gave Victorian writers like Levy—and the other authors discussed in this book—a new understanding of music's material potential and the language to describe its physical and physiological powers. Music not only provides the metaphorical backdrop against which Levy's speaker "cruises the concert hall" but also takes physical shape and substance, inducing bodily pleasures in an entire community of listeners.[14] In acoustical terms, the trembling that the speaker experiences—the music's very presence in their "blood"—reads not as a vague euphemism for sexual arousal but as a concrete and profuse evocation of music's actual effects on the human body. Levy's poem resonates just as much with acoustical theories of sonic nerve stimulation or musically induced blood circulation as it does with fin-de-siècle Decadent poems that euphemize sexual "deviance." When examined through an acoustical lens, Levy's poem represents not simply a coded expression of lesbian desire but rather a much more explosive unsettling of Victorian norms of bodily propriety—an enticing illustration of a delicious, head-to-toe sensual experience shared among a quivering, vibrating "throng."

Reading "Sinfonia Eroica" through an acoustical lens illuminates the poem's depiction of the symphony concert as an erotic and queer event. I use the term *erotic* in the vein of feminist, queer, trans, and asexuality theorists—most famously Audre Lorde and more recently thinkers like Tim Dean and Ela Przybylo—who argue that the "erotic" need not necessarily be limited to genital or even sexual experiences but can refer to, as Lorde writes, any experience or activity (sexual *or* "sensual") that brings an "internal sense of satisfaction," a "lifeforce," or "creative energy empowered."[15] In

Levy's poem, after all, the physical sensations induced by the music extend far beyond genital contact, same-sex desire, or even sex itself. Similarly, I use the term *queer* as it is articulated by theorists such as Sedgwick, Cathy Cohen, Jack Halberstam, Elizabeth Freeman, and Sara Ahmed, who do not limit queerness to the "gender of object choice," as Sedgwick writes, or even to sexuality at all, but understand *queer* as a broader term for modes of living that interrogate the "logics and organizations of community, sexual identity, embodiment, and activity in space and time," to quote Halberstam.[16] The musical space that Levy imagines is queer because it facilitates a desirous gaze between (presumably) two women but also because it makes erotic sensations available to nearly every entity, human and nonhuman, in the concert hall: the speaker, the "throng" of audience members, the music itself, the very air in the room. The speaker derives pleasure from the glimpse of a potential lover as well as from corporeal contact with the music itself, which represents a powerful physical force—a "swell[ing]," "mighty strain" that enters their body and makes them "quiver[]."[17] In the context of new acoustical understandings of music as a material, particle-filled entity with the ability to enter the human body and ignite sensations therein, it is entirely possible that Levy's speaker genuinely does not know "which was sound, and which, O Love, was you."[18] At the risk of making too much of a potential musical pun, the lines "I, with the rest, / Sat there athirst, atremble for the sound" could be read as a reference to a *musical* rest, and thus the formulation of the speaker joining "with" it—awaiting the symphony's start, "athirst, atremble for the sound"—could be seen as another nod to the speaker's visceral connection to, and nervous alignment with, the music.[19] Levy's poem affirms feminist musicologist Suzanne Cusick's famous 1994 inquiry, "What if music IS sex?"—what if music is a source of pleasure and intimacy in its own right, not just a something that is "*like* sex, or [has] the capacity to *represent* sexuality and gender?"[20] For Levy's speaker, sound waves provide as much (if not more) physical intimacy as a sexual partner—a subversive suggestion in a society invested in curtailing women's sexual autonomy and promoting strictly reproductive sex. If music makes for the best kind of lover, what does that mean for the codes that govern Victorian gendered and sexual life?

What it might mean, at least for Levy, is a new framework for eroticism that is not confined to encounters between two bodies. Importantly, it is not just the speaker, nor Sylvia (with her head "lean'd" back), who experiences this erotic musical rapture; it is the entire "expectant throng" of multiply-sexed bodies in the concert hall (the group of "clust'ring heads," "*each man* [who] held his breath").[21] Even the atmosphere seems to get in on the action; the poem repeatedly highlights the physical contact between the music and the air, which is described as "waiting" to be "smote" and "swell'd" upon by sound.[22] Levy imagines the symphony as a queer, utopian space bursting with communal pleasures and reverberating with music that bounces off the walls and seats, permeates the listeners' bodily orifices, and leaves the very air charged with sonic sensation. The music of Beethoven's "Eroica" is thus not a meager replacement for something the speaker would rather be experiencing between the sheets but a vibrant source of erotic contact in and of itself, shared among the entire "thronging" concert hall. As this book shows, Levy was not alone in imagining music's queer, erotic possibilities.

Acoustical Readings

Sounding Bodies argues that nineteenth-century acoustical science enabled some of Victorian literature's most explicit representations of erotic corporeality. At a time when bodies—particularly gendered and sexual bodies—were most often described figuratively in literary texts, acoustical science enabled overt descriptions of bodily affects and sensations. Though acoustical scientists themselves were relatively uninterested in gender or sexual politics, Victorian writers like Levy put acoustical ideas to "queer use," to use Sara Ahmed's term for "how things can be used in ways other than for which they were intended or by those other than for whom they were intended."[23] From realist novelists such as George Eliot and Thomas Hardy to New Woman writers such as Sarah Grand and Bertha Thomas, from creators of fin-de-siècle ghost stories such as Vernon Lee and John Meade Falkner to anonymous authors of underground pornography, a wide range of Victorian writers drew upon acoustical science to depict scenes of music listening and performance as intensely embodied and politically destabilizing events, though ones safely explained and supported by experimental science.

Acoustical science's facilitation of such explicit bodily description had enormous implications for Victorian understandings of gender, sexuality, intimacy, and eroticism. What might it mean, for instance, to have those quivering and pulsing bodies—or those bodies that *produced* the quivering and pulsing music—be *female*, in a world and a literary sphere that rarely granted women such overt moments of physical agency? What might it mean to imagine vibratory musical exchanges between men, or even between groups of people, in a society that virulently insisted on categorizing and demonizing "aberrant" sexualities and kinship structures? What might it mean for a musician to achieve their deepest sense of erotic gratification not from touching another human but from pouring their kinetic energy into their instrument and feeling it resound in response?

The overt corporeality of these musical moments renders them powerful sites for articulations of feminist and queer politics—terms that, while anachronistic in a nineteenth-century context, capture music's interventions in urgent conversations, both then and now, about phenomena such as female agency, gender play, sexual violence, same-sex desire, and nonnormative kinship formations.[24] Victorian writers drew upon the language of sonic sensation to depict and defend the *kinds* of bodies criminalized in their world, particularly those whose gender presentations or preferred forms of intimacy or kinship incited social stigma, legal punishment, and violence and whose overt depiction in literary texts risked censorship or moral scorn. This was especially true for writers marginalized by their own gender identities and sexual practices—protofeminist novelists like Grand and Thomas, or queer writers like Levy and the anonymous authors of *Teleny*—but also for canonical figures like Eliot and Hardy. Music was invisible and intangible yet, in the context of nineteenth-century acoustical science, undeniably palpable—and thus represented the perfect tool through which Victorian writers could imagine new embodied possibilities for their characters' lives. Moreover, the notion, advanced by acoustical scientists, that humans' responses to music are automatic (reflexive and preconscious) and universal (experienced by all living beings) proved politically useful for writers seeking to validate transgressive desires and pleasures as natural and expected rather than degenerate or perverse.

Contemporary scientific understandings of music listening and performance as physical and physiological experiences enabled Victorian writers to depict musical scenes as brimming with a much more capacious range of desires and pleasures than scholars have allowed. In the texts discussed here, female violinists burst onto the stage and activate their arm muscles to produce rapturous sounds that make their audiences writhe; male lovers convulse to each other's piano music; anthropomorphized instruments long for their players' touch; listeners vibrate to the haunting sounds of musical ghosts from previous centuries; and, as in Levy's poem, listeners and players revel in erotic sensations in response to the music itself. Victorian authors used acoustical science to illustrate female bodily power and pleasure, imagine destabilized gender subjectivities, capture the horrors of rape, defend queer sexual desire, and conceive of forms of intimacy outside of nuclear, reproductive kinship structures and even beyond the human world. What makes music such a powerful mode of feminist and queer representation, then, is that it not only gives characters access to gendered and sexual experiences often denied to them but also offers them access to entirely new kinds of erotic sensations altogether. The Victorian writers discussed in this book were onto the ways in which music can provide, as musicologist Jodi Taylor wrote over a century later, "a queer erotic reality beyond the boundaries of gender, sexed bodies, and specific bodily orientations."[25] In the context of acoustical science, musical scenes represent some of the queerest moments in Victorian literature—queer not simply in their descriptions of transgressive gendered or sexual identities and practices but also in their imaginations of broader sets of relations, affinities, and ways of being in the world.

Sounding Bodies locates unexpected feminist and queer possibilities in three famously conservative and exclusionary cultural realms: Victorian literature, nineteenth-century physiological science, and Western classical music.[26] However, my aim is not to ignore the deeply racist and classist histories and traditions of these realms. As Ronjaunee Chatterjee, Alicia Mireles Christoff, and Amy R. Wong argue, scholars of Victorian literature must name the whiteness of their objects of study and critical projects rather than "read[ing] right through them."[27] The Victorian literary texts treated in this book were all written by white British or Irish authors. Their centrality here is a result of what Chatterjee, Christoff, and Wong describe as "the racism that undergirds the history of aesthetics, canon formation, and curricular bias," as well as a symptom of the exclusion of nonwhite people from the Victorian classical music world and thus their erasure from British Victorian literary texts set in musical spheres.[28] Western classical music's notoriously repressive and homogeneous culture means that the modes of resistance charted in these texts, while subversive in some ways, were available only to those with the literal and cultural capital to access musical education, gain permission to play onstage, or accumulate the know-how to obey classical music's stringent codes of etiquette. Moreover, as I discuss further in chapter 1, the physiological discourses used by these authors, including physiological acoustics, were often mobilized for racist and eugenicist ends in the nineteenth century. Thus, while the authors in this book draw upon musical science to articulate what I would deem a feminist and queer politics, these politics were far from fully intersectional. One doubts whether the writers I study would have been as eager to attribute such innate musical abilities and natural musical sensations to bodies that were not, like most of those discussed here, white, British, and middle or upper class. Indeed, the archive suggests they were not. For example, despite the late-century success of Black classical performers such as Amanda Aldridge and Samuel Coleridge-Taylor—two of the first Black musicians to be admitted to the Royal College of Music—to my knowledge, no as-yet-discovered nineteenth-century British fiction portrays Black classical musicians.[29]

The works I discuss are thus ripe for thinking about in the context of Kadji Amin's framework of "deidealization," which he defines as a "form of the reparative that acknowledges messiness and damage" and that acknowledges the unsettling enmeshment of reparation and repression in works often hailed as radical.[30] Texts that offer transgressive imaginings of erotic life do not always realize fully utopian or liberatory representations, and it is dangerous for critics to create a "romance of the alternative" that obscures texts' and authors' other, sometimes more troubling, investments.[31] As Kristin Mahoney, also drawing on Amin, writes in her recent study of queer kinship, it is important to acknowledge phenomena from the past that are "neither entirely radical and ethical nor fully conservative and exploitative" and to attend to the "tension between radical desires and conventional tendencies."[32] My feminist and queer readings of Victorian musical scenes are similarly "deidealized." I do not hail any of these texts as fully (or even mostly) liberatory; indeed, their whiteness, their situatedness in middle- and upper-class realms, and their ties to the exclusionary and often violent spheres of nineteenth-century physiology and classical music preclude such an argument. Rather, I identify moments in which Victorian texts either critique or, to quote Sedgwick, "extract[] sustenance" from their culture.[33] I locate in these works brief glimpses of what the late queer theorist José Esteban Muñoz describes as the "utopian" promises of queerness—to "dream and enact new and better pleasures, other ways of being in the world, and ultimately new worlds."[34] In the texts discussed here, acoustical science creates space for communal forms of queer eroticism too often rendered—like sound itself—invisible.

Below, I outline the book's contributions to several ongoing conversations in the field of Victorian studies: discussions of sound and the senses, music-literature studies, and scholarship on gender, sexuality, and eroticism. *Sounding Bodies* not only teaches readers new things about Victorian literature but also suggests that Victorian literature has much to teach readers about embodied and erotic life. In a twenty-first-century world in which the pay gap between men and women, especially women of color, remains wide; where rape proliferates on college campuses; where abortion is increasingly criminalized; where "eroticism" is still often defined in relation to genital sex, even decades after Lorde's expansion of the term; and where exceedingly narrow frameworks for gendered subjectivities and family structures still dominate, as reflected by the persistence of "gender reveal" parties, the wedding industrial complex, and the state's failure to provide sufficient support for childcare, Victorian gender and sexual mores are not as historically distant as modern-day readers might prefer to think.[35] It is important, even urgent, to look for moments in the past where writers and thinkers were able to imagine alternative possibilities for their characters' erotic lives and occasions for pleasure, intimacy, and community. As scholars continue to think through how (and *why*) Victorian literature should be read and taught, one possibility is that it sometimes offers models, however imperfect, for "other ways of being in the world."

Listening In: Sound, Sensation, and Science in Victorian Studies

Reembodying Victorian Sound Studies

As its title suggests, this book argues that the body should be central to discussions of sound in Victorian literature. Such a corporeal focus illuminates the feminist and queer potential of sonic moments in nineteenth-century texts. *Sounding Bodies* threads together two existing scholarly conversations about Victorian literature and science: studies of acoustics in Victorian literature and studies of physiology and the senses in Victorian literature. Scholars in the field of sound studies have done important work to define the nineteenth century as an "Auscultative Age"—one in which scientists, philosophers, and writers were just as preoccupied with sound as with sight.[36] Literary critics most often focus on sound *physics* in Victorian texts. Gillian Beer, for instance, identifies Gerard Manley Hopkins's fascination with invisible waves of light and sound and his attention to the musicality of language and speech, which were spurred by his readings of Helmholtz and Tyndall.[37] John Picker has demonstrated how Helmholtz's studies of sound waves and pitch frequencies provide rich metaphorical inspiration for Eliot's depiction of the "roar on the other side of silence," Charles Dickens's and George Eliot's illustrations of urban street noise, and Alfred Tennyson's engagement with new sound-recording technologies.[38] Building on this work, I argue that sonic moments in Victorian literature should be read physiologically as well as physically. Victorian acoustical scientists were just as focused on sound's effects on the human body as on its interactions with the air and atmosphere. After all, the subtitle of Helmholtz's landmark 1863 book *On the Sensations of Tone* is A Physiological *Basis on the Theory of Hearing* (emphasis mine). Helmholtz himself noted, "Hitherto it is the *physical* part of the *theory of sound* that has been almost exclusively treated at length. . . . But in addition to a *physical* there is a *physiological theory of acoustics*" of which "not many results have as yet been established with certainty."[39] Helmholtz made it his mission to develop a robust "theory of the sensations of hearing"—one that filtered into Victorian literary texts in ways that scholars have not yet fully explored.[40]

Studies of Victorian literature and acoustical science can thus be brought into productive conversation with studies of Victorian literature and physiology. Literary critics such as Nicholas Dames, William Cohen, and Benjamin Morgan have highlighted the vibrant sensory worlds of Victorian literature, showing how Victorian authors drew on physiological science to illustrate embodied reading practices, describe sensual experiences of sight and touch, incorporate poetic techniques calculated to arouse maximal physical excitement, and create artwork designed to appeal to the human eye.[41] Whereas scholars have tended to focus mainly on the sights, tastes, and smells of Victorian literature, *Sounding Bodies* tunes into its sounds and listens to what they reveal about Victorian writers' engagements with gender, sexuality, and intimacy.

Understanding acoustics as a study of embodied sensation enables feminist and queer readings of Victorian musical scenes. The language and ideas of physiological acoustics were uniquely useful for writers seeking to portray bodily sensations—including pleasure, arousal, and desire—that were otherwise difficult to capture in the nineteenth century for fear of moral offense or censorship. In this context, scenes of music making and listening served as useful sites for authors to delve into the minutest details of their characters' bodies and describe concrete forms of bodily contact. Acoustically, music making and listening are forms of penetration and reception; to play music for another is to enter and stimulate their body. And yet, during musical exchanges, no bodies actually touch; no tactile contact occurs; visually, everything is as it ought to be. The sensations that *do* take place are safely attached to an acoustical-scientific discourse rooted in empirical observation and experimental science—realms that had particular cultural cachet in an era that increasingly prized "objectivity."[42] A violinist's sweating body, for instance, could be traced to contemporary scientific understandings of the physical exertion required to play an instrument—evidence of a biological phenomenon rather than an autonomic response to romantic or sexual attraction. A virtuosic female violinist could be cast not as a dangerous or self-absorbed diva but as an artist whose nerves and muscles are innately primed for high-level performance. A listener's beating heart and quaking limbs at a piano performance could be explained just as easily by the physiology of music as by a sexual attraction to the player.

Moreover, acoustical theories of musical response as an automatic, involuntary, and universal phenomenon rendered it an even more useful tool for Victorian writers to depict gendered and sexual bodies. New Woman novelists, for instance, described at length the visceral powers of their violin-playing heroines, such as their abilities to harness their muscular and nervous strength to make their listeners quiver in response, in order to confront misogynistic Victorian music critics who believed that virtuosic playing was an impossible feat for feeble female bodies. The authors of queer pornography tied same-sex desire to musical response so as to describe both as natural and organic—a powerful intervention at a time when the former was most often thought of as *un*natural. Realist novelists drew upon acoustical understandings of embodied musical response as automatic and uncontrollable to overtly illustrate and critique the harms of gendered violence. That these descriptions were so overt does not, of course, mean that they were categorically "better" or "more radical"; rather, they captured in new ways the phenomenological contours of gendered and sexual life. In the context of Victorian acoustics, music enabled Victorian authors to depict a wide range of bodily sensations and encounters—especially those that were not supposed to happen or that authors were not supposed to describe. While overt portrayals of gendered and sexual bodies were often invisible in Victorian literature, they were sometimes audible.

The Feminist and Queer Uses of Acoustical Science

This book uncovers moments in Victorian literature in which physiological science was accommodating of, rather than hostile to, feminist and queer politics. Literary critics and historians of science most often discuss Victorian physiology as part of Foucault's "medico-sexual regime" that framed "deviant" genders and sexualities as "lesion[s]," "dysfunction[s]," or "symptom[s]."[43] As scholars have long noted, nineteenth-century physiological science was often weaponized against those already marginalized by gender, sexuality, race, class, or ability. Evolutionary and biological science that framed particular kinds of bodies as "naturally" behaving in certain ways often threatened "women's egalitarian aspirations," as Carolyn Burdett writes, or "reduc[ed] queers to mere bodies, passively in thrall to diseased impulses beyond their control," as Dustin Friedman argues.[44] Fraser Riddell outlines how some fin-de-siècle sexologists even drew on the insights of sound science to frame "homosexuals" as "pathologically" sensitive to music.[45] In fact, several of the writers and scientists discussed in this book at times displayed such violent body politics. Sarah Grand was known for her eugenicist beliefs about marriage and reproduction; the authors of *Teleny* cast women as scabby, slimy, diseased, and cadaverous; and both Herbert Spencer and Helmholtz established a sonic hierarchy that, as I discuss in chapter 1, privileged Western music's "natural" tones over "savage" music's rudimentary "noises."[46]

Yet the Victorian writers discussed in this book also, at times, put physiological science to "queer use."[47] *Sounding Bodies* thus resonates with recent work in feminist, queer, and trans science studies that, while still taking seriously the violent potential of essentialist thinking and the exclusionary ways in which scientific knowledge is shaped and produced, argues for renewed attention to the affordances of embodied experience for feminist, queer, antiracist, and anticolonial resistance. Working against what they see as a long tradition of "antibiologism" in gender and queer studies, several theorists have proposed that attending to the body is crucial for registering corporeal experiences of pleasure, desire, pain, and trauma, as well as embodied events like sex, childbirth, breastfeeding, and gender transition.[48] Thinkers like Elizabeth Grosz, Jay Prosser, and Angela Willey have called for, respectively, a "corporeal feminism," a focus on the body's "fleshiness, its nonplasticity, and its nonperformativity," and a "queer feminist critical materialist science studies."[49] The sounding bodies of Victorian literature likewise demonstrate the feminist and queer possibilities that physiological science can open.

Hearing New Things: New Directions for Music-Literature Studies

Music Beyond Metaphor

Sounding Bodies introduces a method for music-literature studies that shifts away from traditional readings of music as a metaphor in literary texts.[50] Much of the existing work in music-literature studies emphasizes what Nina Eidsheim calls "the *figure of sound*"—music as a symbol, metaphor, synecdoche, code, or trope to describe events that can only occur off-the-page.[51] Such figurative interpretations are particularly evident in discussions of music alongside gender and sexuality. This is unsurprising. Even apart from musical contexts, gender and sexuality are often relegated to what Claire Jarvis calls the realm of "fuzzy metaphor," as recent work on "coded erotic scenes," "hidden" abortion plots, "clandestine" representations of marital rape, and concealed pregnancies makes clear.[52] William Cohen argues that the "unspeakability" of sex in Victorian England prompted novelists "to develop an elaborate discourse—richly ambiguous, subtly coded, prolix and polyvalent" to describe things that were "*incapable* of being articulated as well as . . . *prohibited* from articulation."[53] As music-literature scholars have long emphasized, music was one of these "elaborate discourse[s]." When birdsong symbolizes romantic longing, or a heroine's piano skills encode her marriageability and maternal potential, or two lovers' musical exchanges metaphorize the "love that dare not speak its name," music serves as a stand-in "for what cannot be put into words."[54] Drawing on Theodor Adorno's description of music's "indefinite" nature, for example, Joe Law writes that nineteenth-century writers often used music as a "coded reference to same-sex desire."[55]

This work is crucial, to be sure, as it has uncovered important moments of queer representation throughout Victorian literature, such as the insinuations of same-sex love in "Sinfonia Eroica" or the coded descriptions of the murderous "excellent musician" Alan Campbell (a violinist, piano player, and frequent operagoer) in Oscar Wilde's *The Picture of Dorian Gray*.[56] "Coded" representation was, after all, necessitated by the very real dangers of overt representation. Readers need only think of Hardy's lifelong struggles with censorship or Wilde's imprisonment at Reading Gaol, where he was forced to pick fibers out of rope until his fingers bled, to recall the material consequences of overtly describing corporeal life—and especially sexual transgression—in the nineteenth century.[57] For many writers, music represented a crucial tool—even survival strategy—for hinting at experiences whose explicit representation would have been outright dangerous in Victorian England.

However, such "coded" readings often ignore the glimpses that readers sometimes get of characters' phenomenological experiences of music playing and listening. What if Victorian writers were just as interested in music *itself* as a source of erotic sensation? What if readers shift away from interpretations of music as a figurative tool and listen instead for what the actual sounding bodies in nineteenth-century literature have to say? What if scenes of music making and listening in Victorian literature are, above all, about bodies feeling and doing things and interacting with other bodies? Acoustical science propelled Victorian authors to approach music in distinctly less figurative and emphatically *non*-euphemistic ways, rendering scenes of music listening and performance some of the most overtly visceral moments in Victorian literature. It is for this reason that I focus just as much on fiction as on poetry, the latter being the preferred genre of many music-literature scholars. I am interested in concrete descriptions of bodies onstage and in the audience as they play and listen, which unfold most commonly (though not exclusively) in lengthier prose scenes set in concert halls or conservatory studios. This book thus performs what critical musicologists and scholars of music cognition and auditory neuroscience have described as "carnal musicology" or "embodied music theory," which, Cusick writes, offers a "renewed awareness of what is mortal, fleshy, and erotic in our musical pleasures and loves" and, Dana Baitz suggests, invests in the relationship between "musical and bodily materiality."[58] As the nineteenth-century scientists and writers discussed here show, tuning into the musical body—the hands that hold the violin, the fingers that stroke the keys, the spine that tingles in response to a sound wave—can awaken erotic sensations that characters might not otherwise experience or that authors might not otherwise express. For many Victorian writers, music was not merely a way to encode dangerous ideas that needed to remain shrouded in metaphor but also a way to open up their works and make space for a wide range of embodied experiences.

Queering the Concert Hall

By zooming in on actual scenes of music listening and performance in Victorian literature, *Sounding Bodies* highlights the concert hall as an unexpected site for feminist and queer critique. Musicologists and literary critics rightly focus on classical music as one of the most infamous bastions of exclusion and conservatism in Western culture, dominated as it is by "dead white men in wigs."[59] In Victorian England, a society long anxious about its perceived status as the "land without music" compared to the Continent, music critics and practitioners eagerly embraced opportunities to solidify Britain's high-cultural musical status, often by controlling its audiences, training new generations of players and listeners, and defining itself against a series of "others," including bawdy, licentious, and, crucially, lower-class East End music halls, as well as "dangerous" or "demonic" foreign virtuosi.[60]

Victorian music critics dictated rigid rules of comportment for audience members and performers; educators used music drills to discipline schoolchildren, particularly those from lower classes; religious officials used hymns to promote (and sometimes enforce) Protestant ideals; and moralists urged young women to learn piano in order to tame their passions.[61]

It is this picture of Victorian musical culture—dominated by its status insecurities, strict codes of conduct, and anxieties about musical others—that scholars have argued most captured the Victorian literary imagination, particularly in moments when music plays into representations of gender and sexuality. Many literary critics discuss music's imbrication in dominant Victorian gender ideologies. Mary Burgan, for instance, points to the common literary trope of the "heroine at the piano"—the sweet-tempered, gentle female musician who performs solely in the drawing room for her family members and suitors and, in doing so, shores up cherished Victorian ideals of femininity, docility, and domesticity.[62] Phyllis Weliver examines literary constructions of musical women as either "angels" or "demons" (and often a blurring of both).[63] Alisa Clapp-Itnyre argues that music was at once a "corrective to foster patriotism, morality, spirituality, and domestic tranquility" and a dangerous source of " 'artificial' display" and " 'immoral' sensuality."[64] Moreover, as a result of Victorian sexological writing that disparaged the musical affinities of "homosexuals" as "pathological," Riddell argues, queer writers often made efforts to distance music from the body, "refusing an embodied materiality that taints musical experience with sexual abnormality."[65]

However, this book reveals a set of authors who saw embodied music making and listening as opportunities to imagine gendered and sexual subversion. A musical performance was not only a time to sit still and be silent but also a moment where one's every nerve could be ignited. If, as acoustical science explained, classical stages and concert halls were full of vibrating bodies, then weren't they always already queer spaces?

Good Vibrations: Expanding the Erotic in Victorian Studies

Sounding Bodies expands current scholarly understandings of the queer, erotic contours of Victorian musical representation—and of Victorian literature in general. Once readers see Victorian musical scenes as brimming with all kinds of vibrating bodies, a range of queer, erotic relations come into view—between performers and listeners, ensembles and audiences, players and instruments, and humans and music itself. While sometimes sexual, as when the protagonist of *Teleny* experiences an orgasm in response to his lover's piano playing, the pleasures of music do not always involve genital or even sexual arousal but are nonetheless intensely corporeal—inducing full-bodied sensations of arousal and stimulation. In this context, queer musical erotics are just as likely to appear in canonical realist novels as in fin-de-siècle pornography. The Victorian texts I discuss can thus be said to anticipate efforts by twentieth- and twenty-first-century feminist, queer, and asexuality theorists to expand the definition of the "erotic," as discussed earlier.[66] In their portrayals of listening and hearing as fundamentally corporeal experiences shared among collections of human and nonhuman bodies, these works conceive of what Przybylo describes as "erotic . . . forms of intimacy that are simply not reducible to sex and sexuality and that challenge the Freudian doxa that the sexual is at the base of all things."[67]

Scholars of Victorian literature and culture have for decades prioritized locating gender and sexual identities, social relations, and forms of kinship that challenge stereotypes of the period as unequivocally prudish and conservative. For example, Lisa Hager, Simon Joyce, and Ardele Haefele-Thomas underscore how Victorians "negotiated the possibilities of gender diversity well before sexologists invented the clinical term 'transgender,'" as Haefele-Thomas writes.[68] Resonating with claims by queer theorists that "queer" does not always mean "antinormative," Victorianist scholars have also emphasized that queerness was often hiding in plain sight in nineteenth-century literature and culture—recognized, acknowledged, overtly discussed, and sometimes sanctioned by mainstream society and even found in canonical novels.[69] Critics such as Deborah Lutz, Ellen Bayuk Rosenman, Sharon Marcus, Holly Furneaux, Abigail Joseph, Talia Schaffer, Dustin Friedman, and Kristin Mahoney have identified scenes of male nursing, sibling intimacy, female friendship, religious interaction, medical examination, textual exchange, and artistic brotherhood as sites of eroticism that, while not necessarily genital or even sexual, are nonetheless corporeal, intimate, pleasurable, and sustaining.[70] *Sounding Bodies* adds music players and listeners to this growing list. Not only is *Teleny*'s musical orgasm erotic, but so is the energy a female violinist gathers while performing, the arousal that audience members feel while attending a concert, and the tactile intimacy a lonely musician finds by fingering her piano keys. It is hard to imagine a more vivid moment of erotic bodily gratification, for instance, than in Mary Augusta Ward's *Robert Elsmere*, in which a woman feels her "strange little veins of sentiment running all about her" as she hears the protagonist play her violin, or in M.

E. Francis's *The Duenna of a Genius*, when the heroine attends a concert at St. James's Hall and eagerly and automatically fingers along with the violin passages, "her hands working, the fingers curving themselves involuntarily, as though they too itched to handle bow and strings."[71]

The works discussed here present fictional instantiations of Cusick's landmark proposal ("what if music IS sex?") and the feminist and queer musicological threads that followed, including Taylor's notion of "aural sex" and further discussions by Cusick, Judith Peraino, and Fred Everett Maus of music as an erotic experience.[72] Note the striking resemblance between Levy's "Sinfonia Eroica" and Cusick's description of her erotic experience at the Metropolitan Opera in 2013:

> I felt Lorde's erotic energy in the intimacy I shared with singers whose vocal intimacy with each other was enacted hundreds of yards away. . . . If I listened with eyes closed or averted there was no distance, only the sounding intimacy all around me. That intimacy was in (and with) the vibrating air, wood, gold leaf, plush, and the four thousand human bodies who listened with me, as much as it was between the two singers on that distant stage. All of us, animate and inanimate, were in an intimate contact with each other that preserved yet overflowed our differences.[73]

Eroticism can be found in making music for or with another, in touching one's instrument and having it reverberate in turn, in sensing the electric charge of being part of an audience, or in being enveloped by pulses of vibrating air. All one has to do is listen.

Program Notes: Plan of the Book

Sounding Bodies surveys the "queer use[s]" of acoustical theory by a wide range of Victorian authors. The first chapter lays the historical groundwork for this study by chronicling the emergence of acoustical science and the intellectual debates it sparked in the nineteenth century. Acoustical science was controversial, as it unsettled long-held Romantic and idealist philosophies that hailed music as a transcendent phenomenon whose mysteries needed to be preserved. These new acoustical theories also undermined cherished humanist notions of the musical genius, the sophisticated listener, and even the autonomous, conscious individual. Contested as it was, however, acoustical science filtered into almost every sector of Victorian culture, capturing the imaginations of musicians, philosophers, educators, politicians, physicians, and writers and compelling them to consider anew the visceral aspects of aesthetic experience.

Each subsequent chapter of the book highlights a set of Victorian authors who used acoustical science to articulate feminist and/or queer politics. The chapters are organized thematically rather than chronologically, with the focus of each shifting from gender (chapters 2 and 3), to sexuality (chapters 4 and 5), to broader forms of eroticism beyond the strictly sexual realm (chapters 6 and 7). Over the course of the book, the queer possibilities afforded by music become more and more capacious. Chapter 2 focuses on a pair of late-Victorian novels—Mary Augusta Ward's *Robert Elsmere* (1880) and M. E. Francis's *The Duenna of a Genius* (1898)—that imagined a rather rare figure in Victorian literature and culture: the professional female violinist. While most musical women in the nineteenth century resembled Burgan's "heroine at the piano," the "lady violinists" in Ward's and Francis's texts pursue professional ambitions, access high-level musical training, perform in public, and garner worldwide acclaim.[74] In an environment where string playing by women was often deemed either inappropriate or impossible, the heroines of these novels perform with intense physical and physiological power, their music reverberating through large spaces and driving audiences to sweat and quiver. Acoustical science enabled Ward and Francis to cast the musical pursuits of their heroines not only as gratifying and pleasurable but also as empirically explained and thus defensible to a Victorian public deeply suspicious of musical women.

While chapter 2 centers on literary articulations of strong and stable female subjectivities, chapter 3 is more concerned with destabilized gender identities. Sarah Grand's *The Heavenly Twins* (1893) and Bertha Thomas's *The Violin-Player* (1880) feature female performers who do not always perform *as* women; both heroines repeatedly cross-dress to gain access to all-male musical spheres. These novels thus reveal a compelling tension between the performance of music and the performance of gender. While the heroines' cross-dressing destabilizes their gender identities, the references to acoustical science in these scenes continually recall the presence of the body. Ward and Francis celebrate the specifically *female* musician, whereas Grand and Thomas prioritize their heroines' musicality above their womanhood. At a time when women's deepest gratification was supposed to derive from their patently feminine roles (marital sex, motherhood), Grand and Thomas find an alternative in cross-dressed violin playing.

Though the works treated in chapters 2 and 3 gesture toward new and promising possibilities for their characters' gendered lives, chapter 4 examines how rapturous music can facilitate sexual violence. Focusing on Eliot's *The Mill on the Floss* (1860) and Hardy's *Desperate Remedies* (1871) and "The Fiddler of the Reels" (1893), I argue that both authors foreground the physical harm that occurs when male players use their music to violate the bodies of their female listeners. I refer to these encounters as musical rapes, drawing on the work of feminist rape theorists such as Erin Spampinato, who argues for a more "capacious conception of rape" that is not limited to forms of nonconsensual genital penetration.[75] Understanding these violent musical encounters as nothing less than instances of rape brings into focus Eliot's and Hardy's profound—and strikingly explicit—explorations of gendered and sexual violence.

While chapter 4 focuses on sexual practices that *should* have been criminalized but were not, chapter 5 focuses on those that *were* criminalized but should not have been. Chapter 5 examines the anonymously published pornographic novel *Teleny* (1893), which famously opens with the protagonist, Camille Des Grieux, experiencing an orgasm in a Parisian concert hall in response to the piano music of his future lover, René Teleny. By suffusing the depictions of their relationship with acoustical language, the authors cast same-sex desire as natural and empirically explained—patently *not* a "crime against nature," as the law suggested.[76] The authors of *Teleny* invite readers to ask: If the tingles of a lover's touch approximate the quivers induced by the sounds of a sonata—ones that everyone in the concert hall experiences—how "diseased" can they really be? In *Teleny*, then, music serves not as a code or symbol for hidden desires but rather as a tool to defend "deviant" sexual lives. I propose that *Teleny*'s queer narrative also extends beyond its depiction of same-sex desire to encompass the erotic potential of humans' interactions with music itself. After all, Des Grieux, Teleny, and the other audience members are just as aroused by sound as by sex. The authors of *Teleny* imagine eroticism as something that can occur beyond the realm of strictly human-human encounters.

Chapters 6 and 7 pull further on these nonhuman threads by exploring texts that depict forms of intimacy between humans and their instruments (chapter 6) and between humans and ghosts (chapter 7). Though these interactions are not always strictly sexual, they are still deeply erotic and thus offer further glimpses into the vast range of embodied experiences explained by nineteenth-century acoustical science. Chapter 6 focuses on human-instrument intimacies in four Victorian texts: Dickens's *Dombey and*

Son (1848), Eliot's *The Mill on the Floss*, Anthony Trollope's *The Warden* (1855), and Hardy's "Haunting Fingers" (1928). While scholars often read musical instruments as conduits for human-human relations, these works depict instrument playing as an acutely gratifying experience in its own right, one that grants both the musicians and their instruments powerful sensations of bodily pleasure and sustaining experiences of intimacy, eroticism, and kinship. In their depictions of human-instrument encounters, then, the authors discussed here imagine erotic pleasure as something available not just between lovers but also between fingers and keys, arms and bows, and skin and resin.

The final chapter takes the discussion of musical erotics even further beyond the human world and into the realm of the supernatural. Chapter 7 considers two fin-de-siècle ghost stories—John Mead Falkner's novella *The Lost Stradivarius* (1895) and Vernon Lee's short story "A Wicked Voice" (1890)—in which young men develop erotic attractions to the music of ghosts who return from previous centuries to haunt them. Both authors use the language of physiological acoustics to cast the ghosts' supernatural music as paradoxically material, exerting palpable effects on the human protagonists, who throb, shake, sweat, and moan in response to the sounds they hear. While critics often discuss ghost stories as representations of queer absence or loss, I argue that Falkner and Lee depict musical hauntings as encounters that provide their protagonists with bodily pleasures and intimacies otherwise inaccessible to them. Moreover, as ghosts are, by definition, figures from the past, these musical hauntings offer forms of intimate contact with history itself in ways that resonate with twenty-first-century theories of queer temporality, such as Elizabeth Freeman's notion of "erotohistoriography."[77] These musical ghost stories reveal that aesthetic experience—and its attendant embodied pleasures—can exist far beyond the land of the living, vibrating across continents and centuries.

Although the story of *Sounding Bodies* unfolds mainly in the nineteenth century, its discussions of music and embodiment reverberate with urgent concerns in today's classical music world. Western classical music culture, after all, still often insists on erasing any traces of the bodies of players and listeners by dressing performers in concert black, silencing audience members' coughs and claps, and systemically excluding bodies that are not white, male, or abled. As I propose in the brief coda, contemporary classical music culture could take a cue from the Victorian scientists and writers discussed in this book, who understood that the powers and pleasures of a musical performance stem from music's ties to the body. After all, as the

Victorian writers featured here demonstrate, even the most sublime sounds are produced by live bodies, ones that move and breathe with the music and whose muscles and nerves create tones that excite the senses and can, at least for a moment, erase the "world's great woe."

Part One

Sounds and Bodies

Chapter One

Hearing, Touching, Feeling Sound
Acoustical Science in Nineteenth-Century Britain

At his Christmas Lecture at the Royal Institution on December 27, 1873, John Tyndall enraptured his audience members with demonstrations of the wonder of sound. Inviting children to sit in the front of the theatre (the lecture was "prepared specifically for them"), Tyndall presented a variety of acoustical toys and tricks, from "magic wands"—rods he made to vibrate in response to musical notes—to "singing flames"—gaseous emissions that "danced" when placed near sounding bells (see fig. 1.1).[1] The *Daily News* reported on the audience's enthusiasm for Tyndall's theatrics: "The beaming faces that circled the stage at Covent Garden or Drury Lane on Friday night did not outshine in signs of cheerful appreciation of the performance those which surrounded Professor Tyndall in the Lecture Hall of the Royal Institution on Saturday afternoon."[2]

The marvels of Tyndall's "performance" owed to the physical and physiological properties of sound. His magic wands and singing flames revealed sound's ability to travel through the air and vibrate physical objects and substances. Tyndall also used the magic wands as models for his explanation of the human body's response to sound; when sound waves come into contact with the human ear, he explained, the ear vibrates just like the magic wand: "This wave impinging upon the tympanic membrane causes it to shiver; its tremors are transmitted to the auditory nerve, and along the auditory nerve to the brain, where it announces itself as sound."[3] Though the *Daily News* reporter found this language of "tremor[s]" and "shiver[s]" rather "an alarming proposition, especially when set forth for the benefit of a juvenile

Figure 1.1. Valentine for John Tyndall, "Sonorous Sensitive!!!!," nineteenth century, English School, pen and ink, from the Royal Institution, UK, London. *Source:* © Royal Institution/Bridgeman Images. Used with permission.

audience," Tyndall's lecture nonetheless induced "joyful chirruping" among all in attendance.[4] Tyndall delighted his listeners with the suggestion that the world around them was vibrating with invisible, yet patently physical, waves of sound, which would travel along their nerves and make their bodies shake and shiver.

This account of Tyndall's Christmas Lecture reflects the burgeoning nineteenth-century fascination with the materiality of sound—and in particular music, the type of sound that, acousticians theorized, vibrated most consistently and thus most powerfully.[5] Nineteenth-century acoustical science was dominated by two specific branches of study, which Helmholtz called "physical acoustics" and "physiological acoustics."[6] Whereas physical acoustics revealed new ideas about sound waves, frequencies, and vibrations, physiological acoustics centered on sound's interactions with the human ear and with the nerves and muscles of the rest of the body. Physical and physiological acoustics were so controversial—and so compelling—because they invited people to attend to the vibrating physical world around them, to consider the minutest functions of their nervous systems, to question their bodily autonomy, and to reconceptualize their relationships to art. If hearing also meant physically feeling, even touching, then what new possibilities could humans imagine for their aesthetic and embodied lives?

Physical Acoustics

In the late 1780s, the German natural philosopher Ernst Chladni conducted an unusual experiment involving glass, sand, and a violin bow. Chladni gathered a collection of glass plates and sprinkled handfuls of sand on their surfaces. He then drew a violin bow across the sides of the glass plates, which caused the sand to shake into distinct, orderly, and even beautiful patterns that altered depending on the force and speed of the bow strokes (see fig. 1.2).[7] Through these experiments, Chladni, whom Tyndall later deemed the "father of modern acoustics," made the invisible visible.[8] His sand-covered plates provided a visual record of something as yet unseen: the material effects of sound.

Scientific interest in physical acoustics increased in the mid- to late eighteenth and early nineteenth centuries. Following decades of "ocularcentric" scientific inquiries, scientists began to shift their attention from sight to sound, investigating whether the physics of light transmission and color production that explained human vision could also explain the process of

28 | Sounding Bodies

Figure 1.2. Illustration of Chladni plates, 1867, in John Tyndall, *Sound*, 1st ed. (London: Longmans, Green, and Co., 1867), 142. *Source:* HathiTrust Digital Library. Public domain.

sound production.[9] Experiments such as Chladni's confirmed what scientists had long hypothesized: that sound was caused by vibrations in the air. This concept took firm hold in the scientific community as the century progressed.[10] In early nineteenth-century England, scientists like Charles Wheatstone, David Brewster, Michael Faraday, and John Herschel popularized Chladni's discoveries and developed their own acoustical theories, presenting them at venues ranging from the Royal Society and Royal Institution to Pall Mall showrooms and shops.[11]

By the mid-nineteenth century, Helmholtz had turned his attention from his well-known studies of optics to the sonic phenomena that Chladni and others described.[12] Helmholtz believed that the science of music had not yet received in-depth scientific treatment—a fact he attributed to music's invisibility and intangibility.[13] Spurred by the growing prestige of the physical sciences and an "expanding laboratory culture" in Germany and across Europe, in 1856 Helmholtz began to conduct his own acoustical experiments and give popular lectures on sound, including a series at the Royal Institution in April 1861.[14] These investigations culminated in his landmark work *On the Sensations of Tone as a Physiological Basis for the*

Theory of Music (1863), which he hoped would be "popular and generally understandable . . . accessible to musicians and music aficionados as well as to physiologists and medical doctors."[15] Indeed, *Sensations* was wildly successful and was translated into English by the British scientist Alexander Ellis in 1875 and again in 1885.[16]

One of Helmholtz's major interventions in physical acoustics was his distinction between music and noise. Though music and noise could "intermingle in very various degrees, and pass insensibly into one another," Helmholtz wrote, musical notes arose from very specific physical conditions.[17] While noise resulted from a "rapid alternation of different kinds of sensations of sound . . . tumbled about in confusion," music derived from regular, rapid, and periodic vibrations of the air.[18] Tyndall later devised an especially playful description of this phenomenon: if the "puffs of a locomotive" accelerated rapidly enough, "the approach of the engine would be heralded by an organ peal of tremendous power."[19]

Helmholtz's most famous claims in *Sensations* concerned his theory of "sympathetic vibration" or "sympathetic resonance."[20] According to Helmholtz, when a "sounding body" emitted a musical tone, the resulting vibrations could cause a nearby "sympathising body" (such as a string, bell, piece of glass, membrane, or other "elastic" body) to vibrate in response.[21] Helmholtz said this occurred when the vibrations of the "sounding body" corresponded to, or oscillated at the same periodic time as, those of the "sympathising body."[22] Sympathetic vibration explains why Chladni's sand shook on the plates, why opera singers' voices can break glass, and why, in popular legend, Beethoven's floor trembled when he played the piano with especial vigor.[23] In one of his many demonstrations of sympathetic vibration, Helmholtz employed a voice teacher named Emma Seiler to sing loudly into a piano to see if the strings would vibrate intensely enough to move the papers Helmholtz had placed upon them.[24]

Though Helmholtz made many friends among the "British scientific elite," Tyndall was perhaps Helmholtz's most enthusiastic follower in the field of physical acoustics.[25] Tyndall met Helmholtz in 1861, and the two men spent much time together, even attending "gymnastic performances at the Alhambra" with other British scientists like George Busk, John Lubbock, and William Carpenter.[26] Tyndall focused many of his own Royal Institution lectures on physical acoustics and collected eight of them into his 1867 book *Sound*.[27] In an effort to "express Helmholtz's views in the briefest possible language," Tyndall developed a series of simple visual models to describe precisely *how* sound traveled through the air. Using a row of glass balls that transferred motion to each other when moved, for instance, Tyndall showed that sound was conveyed through the air "from particle to particle" (see fig. 1.3).[28] In what he called another "homely but useful illustration," Tyndall lined up his assistants and used their bodies to visualize the process of sound propagation (see fig. 1.4): "I suddenly push A, A pushes B, and regains his upright position; B pushes C; C pushes D; D pushes E; each boy after the transmission of the push, becoming himself erect. E, having nobody in front, is thrown forward. . . . Thus, also, we send sound through the air, and shake the drum of a distant ear, while each particular particle of the air concerned in the transmission of the pulse makes only a small oscillation."[29]

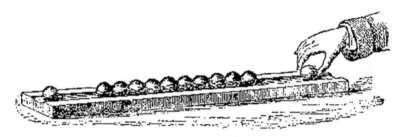

Figure 1.3. Tyndall's glass ball demonstration, 1867, in John Tyndall, *Sound*, 1st ed. (London: Longmans, Green, and Co., 1867), 3. *Source:* HathiTrust Digital Library. Public domain.

Figure 1.4. Tyndall's human sound propagation diagram, 1867, in John Tyndall, *Sound*, 1st ed. (London: Longmans, Green, and Co., 1867), 4. *Source:* HathiTrust Digital Library. Public domain.

For nineteenth-century acoustical physicists, then, musical notes were not evanescent tones from the heavens but material particles that shook the air at particular speeds, vibrating the physical world through which they traveled. As scientists would soon discover, what music did to Chladni's glass plates, Helmholtz's piano strings, and Tyndall's glass balls, it could also do to the human body.

Physiological Acoustics

For centuries, philosophers and scientists have obliquely tied music to the body, using images of musical strings, keys, and pipes as metaphors or models for the spirits and passions, as seen in the Neoplatonic notion of *musica humana* or the figure of the aeolian harp.[30] In the late eighteenth century, anatomists and associationist thinkers frequently used images of musical strings to model how vibrations were transmitted to the soul or brain, via either the "animal spirits" or, as scientists like David Hartley speculated, the nerves (a concept that, according to Shelley Trower, Hartley understood more theoretically than concretely).[31] Enlightenment thinkers like Edmund Burke discussed how certain kinds of music could awaken particular feelings or passions in the human body; long, strong, shrill, and harsh tones, for example, could "excite mirth, or other sudden and tumultuous passions," while slow, soft, and smooth sounds induced feelings of "languor."[32]

In the mid-nineteenth century, discussions of music and the body began to preoccupy evolutionary scientists such as Herbert Spencer and Charles Darwin, who sought to understand why humans made—and were so moved by—music. In his essay "On the Origin and Function of Music" (1857), Spencer theorized that singing resulted from emotionally charged speech; mental excitement could produce "muscular excitement" that affected the "loudness, timbre, pitch, intervals, and rate of variation" of the voice.[33] For singers, Spencer wrote, the muscles of the chest, larynx, and vocal cords expanded and contracted "in proportion to the intensity of the feelings."[34] Though Darwin famously disagreed with Spencer's "speech theory"—he insisted that speech derived from music, not vice versa—he, like Spencer, also saw music as inextricable from the body.[35] In *The Descent of Man* (1871), Darwin described sexual selection as the basis of music: animals produced pleasing sounds in order to attract mates and reproduce.[36] For both Spencer and Darwin, then, the "origin and function" of music had everything to do with corporeality.

Acoustical scientists like Helmholtz and Tyndall brought investigations of music and the body into less speculative, more empirical realms. Supported by a midcentury "expanding laboratory culture," they set out to measure and record the effects of music and its impact on the muscular and nervous systems.[37] For them, music was not simply a model *for* the nerves as it was for Hartley and Burke; it was an actual, physical entity that interacted *with* the nerves in quantifiable ways.

Helmholtz's and Tyndall's physiological studies of hearing owed to decades of research on auricular anatomy. Though anatomists had studied the structure of the human ear for centuries, sound scientists at first struggled to deploy these anatomical observations to explore the mechanisms of human hearing. In his 1855 history of auditory science, the aurist William Robert Wilde (Oscar Wilde's father) wrote that the field was still "at a very low ebb, particularly in Great Britain" at the beginning of the nineteenth century.[38] This "low ebb," Jonathan Sterne has suggested, was due in part to the anatomical complexities of the ear, as scientists struggled to untangle how the dozens of tiny, intricate bones and membranes inside the organ interacted in the process of hearing.[39]

The earliest physiological studies of hearing were spurred by augmenting concerns about hearing impairments and deafness in the nineteenth-century British public, caused by diseases such as smallpox, scarlet fever, and measles, as well as urban lifestyle factors like industrial noise.[40] Studies of deafness helped scientists apply the theories of sound vibration introduced by Chladni to the process of hearing.[41] In 1816, the surgeon and anatomist John Harrison Curtis founded the Royal Dispensary for Diseases of the Ear in Soho Square to study the physiological mechanisms of hearing loss, which he discovered could result from phenomena such as the accumulation of cerumen (earwax) that deadened sonic vibrations.[42] This led him to theorize that sonic vibrations could be used to help people with hearing loss sense sound through physical touch.[43] Curtis experimented with a variety of acoustical instruments, including the hearing trumpet, the artificial ear, and his own invention: the acoustic chair, an armchair with a sound barrel affixed to either side that would amplify nearby sounds (see figs. 1.5 and 1.6).[44] He discovered that when wires were connected between patients' mouths and ears, sound vibrations traveled up their eustachian tubes and enabled them to feel sound in their throats.[45] Such discoveries about sound's ability to vibrate the body would prove essential to treatments of deafness for decades. In her 1903 memoir *The Story of My Life*, for instance, Helen Keller described touching a speaker's throat or laying her hands on a piano while it was being played to feel the sonic vibrations.[46]

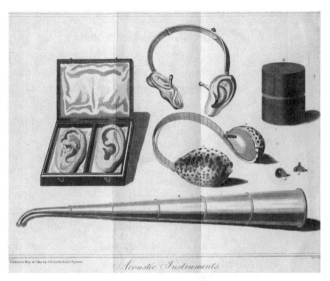

Figure 1.5. "Acoustic Instruments," 1819, in John Harrison Curtis, *A Treatise on the Physiology and Diseases of the Ear*, 2nd ed. (London: John Anderson, 1819), n.p. *Source:* Wellcome Collection digital collections. Public domain.

Figure 1.6. Curtis's acoustic chair, 1842, in John Harrison Curtis, *On the Cephaloscope and Its Uses in The Discrimination of the Normal and Abnormal Sounds in the Organ of Hearing* (London: John Churchill, 1842), 49. *Source:* Archive.org. Public domain.

Despite these developments, by 1864, Helmholtz believed that the scientific world still had relatively little understanding of the "physiological aspects of hearing."[47] Though, as Steege notes, Helmholtz never adopted a fully materialist view of hearing—he reassured his "metaphysico-esthetical opponents" that he did not "undervalue[] artistic emotions of the human mind"—he was nonetheless committed to establishing "the physiological facts on which esthetic feeling is based."[48] Helmholtz knew that music was an art of the senses; he just needed to figure out exactly how it worked.

Helmholtz's research in physiological acoustics was indebted to the earlier work of his teacher and mentor in Berlin, the German anatomist and physiologist Johannes Müller. Müller was famous for his theory of "specific nerve energies," which argued that each sense organ responds differently to external stimuli.[49] To describe his doctrine, Müller drew on the age-old analogy between human body and musical instrument, likening the brain to a "many-stringed instrument, whose strings resound as the keys are touched."[50] Spurred by Müller's theory of nerves and the senses (and, as David Cahan chronicles, his firm but supportive mentorship), as well as the growing popularity of "reductionist physiology" and "organic physics" among Helmholtz's friends at the Berlin Physikalische Gesellschaft, Helmholtz delved into a series of experiments on the anatomy of the ear and the process of hearing.[51]

Aided by the Parisian instrument maker Rudolph Koenig, and expanding on earlier theories of resonance by thinkers like Georg Ohm, Max Müller, and Alfonso Corti, Helmholtz built a device called a resonator, a spherical object made of glass or brass that fit into the ear at one end and connected to a sound source on the other (see fig. 1.7).[52] Each resonator was tuned to a particular frequency and thus allowed Helmholtz to isolate specific musical tones, particularly upper partials (also called overtones), and measure their capacity to produce "sensations of tone" in the human ear. It was through these resonator experiments that Helmholtz determined that "sympathetic vibration" occurred not only in physical objects but also in the human body.[53] Thus, while Helmholtz was far from the first scientist to discover that objects vibrate in response to sound—thinkers as early as Galileo experimented with the effects of pitch on glass and string—he was, according to Picker, "the first to place it so centrally and with such lucid precision in a broadly conceived theory of hearing."[54]

Helmholtz outlined the ear's capacity for sympathetic vibration in detail. Combining the insights gleaned from his resonator experiments with recent "microscopic discoveries respecting the internal construction of the

Hearing, Touching, Feeling Sound | 35

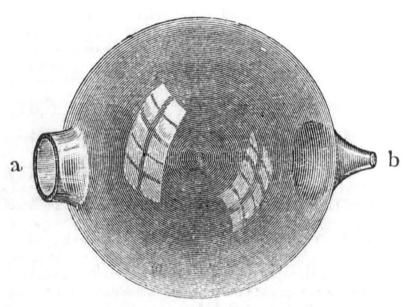

Figure 1.7. Helmholtz resonator, 1863, in Hermann von Helmholtz, *On the Sensations of Tone*, 3rd ed. (London: Longmans, Green, and Co., 1895), 43. Source: Archive.org. Public domain.

ear," Helmholtz concluded that the ends of the auditory nerve fibers were connected to "small elastic parts" that were "set in sympathetic vibration by the waves of sound" and then "excite[d] the nerves by their vibrations."[55] Concretizing the centuries-old metaphor of body as stringed instrument, Helmholtz urged readers to imagine that each "appendage" of the ear was "tuned to a certain tone like the strings of a piano," so that "when . . . that tone is sounded, the corresponding hair-like appendage may vibrate, and the corresponding nerve-fibre experience a sensation."[56] For Helmholtz, almost every part of the ear played a role in this vibratory process of hearing. For instance, the "tympanum," or eardrum—located within the fluid-filled cavity (or "labyrinth") of the ear—worked to "conduct the vibrations of the air with sufficient force."[57] The labyrinth also contained Corti's organ, which Tyndall described as a "lute of 3,000 strings" that "accept[s] the music of the outer world, and render[s] it fit for reception by the brain."[58] According to Helmholtz, these varied and complex processes of vibration and transmission determined how "human beings and mammals" alike perceived musical sound; differences in pitch and tone quality were determined by the "difference in the fibres of the nerves receiving the sensation."[59] As Steege writes, the power of Helmholtz's intervention was that it transformed earlier "abstract" models of hearing into theories that offered "a vivid sense of [its] raw corporeality."[60]

As Helmholtz's ideas proliferated throughout late nineteenth-century Britain through his own lectures at the Royal Institution and Alexander Ellis's translations of *Sensations*, scientists continued to develop physiological theories of music, focusing especially on music's effects on other parts of the body beyond the ear. Helmholtz himself had considered the sensitiveness of other parts of the body to sympathetic vibrations; he wrote of the tympanum's connection to the throat through the eustachian tube and noted that deaf people could "perceive the motion of the air which we call sound" on the surface of the skin, where the "vibration of elastic bodies heard by the ear" could create a "whirring fluttering sensation."[61] Expanding on Helmholtz's work, Edmund Gurney, a British scientist who took Darwin's side in the musical debate with Spencer, wrote at length about the abilities of certain kinds of music, particularly music with strong and constant rhythmic impulses, to induce kinesthetic motion in the muscles: "In melody . . . there is perpetually involved something more even than a suggestion of movement, namely, a direct impulse to move; which is not only felt but constantly yielded to in varying degrees."[62] Gurney continued, "Owing to the physically stimulating power of musical sound and the extreme distinctness and determinateness of the physical sense of rhythmic motion, our corporeal life is brought before us in the most direct and striking manner."[63] The French physician Hector Chomet shared Gurney's interest in muscular responses to music; in his treatise *The Influence of Music on Health and Life* (1874), Chomet recounted how playing piano led him to experience "spasmodic movements of the throat and muscles, when playing certain strains."[64]

Theories of vibratory sympathy, scientists soon discovered, could also be applied to the circulatory and respiratory systems. Chomet described a rudimentary experiment in which a French composer named André Grétry placed his fingers on an artery in his arm and began to sing at different tempi; he felt his pulse "quickening or slackening its action to accommodate itself by degrees to the *tempo* of the new air."[65] In the 1890s, the Russian physician Alexandre Dogiel used a device called a peltismograph to measure the circulation of the blood in response to musical sounds.[66] An 1894 article titled "Physiological Effects of Music" reported the findings of a "Russian physician" (likely Dogiel):

> An influence on the circulation of the blood is noticed, the pressure sometimes rising and sometimes falling, though the action of musical tones, and pipes both on animals and men expresses itself, for the most part, by increased frequency of the beats of the heart; the variations in the circulation consequent upon musical sounds coincide with changes in the breathing, though they may also be observed quite independently of it; the variations in the blood pressure are dependent on the pitch and loudness of the sound and on tone colour.[67]

Music not only tickled the skin and quivered the muscles but also affected the blood, pulse, heart, and breath.

Some scientists described music's effects on more unexpected parts of the human body. Rodolphe Radau's book *Wonders of Acoustics* (1886) detailed humans' ability to "hear through the teeth" and noted that deaf people "have even been known to hear by the epigastrium" (the upper abdomen).[68] The sexologist Havelock Ellis believed that the stimulating effects of music (particularly "orchestral music") could extend to the genitals, resulting in "definite sexual excitement" in many "normal educated women" and in many "neuropathic subjects" (though only in a "small proportion of men").[69] He described a "very beautiful, but highly neurotic, woman" for whom "the hearing of beautiful music, or at times the excitement of her own singing, will sometimes cause intense orgasm."[70] No part of the human body was untouched by the power of music.

"Not Mere Vibrations": Aesthetic Debates

Physical and physiological acoustics were hotly contested in Victorian culture. In the musical world, many still heralded late eighteenth- and early nineteenth-century Romantic and idealist conceptions of music that focused on its metaphysical and spiritual (rather than material and scientific) status.[71] While, as mentioned above, earlier eighteenth-century associationist thinkers were actually quite interested in speculating about music's effects on the body, later eighteenth- and nineteenth-century aesthetic philosophers "relegate[d] the body to a marginal position" in musical writings.[72] The German music critic E. T. A. Hoffmann's 1810 essay on Beethoven's Fifth Symphony captures this Romantic ethos: "Music's sole object is the expression of the infinite.... Music discloses to man an unknown kingdom, a world having nothing in common with the external sensual world which surrounds him."[73] As fully instrumental music, or "absolute music," became a dominant genre in the nineteenth century, many critics prized music's potential to offer an escape from the earthly world of the senses.[74]

These Romantic musical ideals persisted through the late nineteenth and early twentieth centuries, as composers like Gustav Mahler and Richard Wagner continued to idealize what Alex Ross calls the "cult of the Work," which valued the musical composition above the performance, the performer, or the audience.[75] Conductors and music directors during this time endeavored to curate the concert hall as a space devoid of reminders of anything that would threaten "pure" musical contemplation.[76] They incorporated dimmable hall lights to signal audiences to fall "into rapt silence," made concert hall seats more comfortable to prevent listeners from shuffling, and created smaller boxes to prevent social gatherings.[77] Manuals for concert etiquette abounded, prohibiting, as one 1864 manual read, "talking, laughing, fan-gyrating . . . lobbying . . . programme crumpling, and chair shuffling" during concerts—all of which "lowe[r] the Art."[78]

Unsurprisingly, then, the nineteenth-century musical world did not entirely embrace scientific theories that cast music as a patently physical and physiological art form. In 1906, the music critic Joseph Goddard wrote that the ideas of Spencer, Darwin, and Helmholtz threatened to "attack[]" musical aesthetics, as their works failed to provide "an exhausting and convincing philosophy of the higher beauty of music."[79] Even Mary Somerville, the "Queen of Nineteenth-Century Science," worried that materialist theories of music were incompatible with her beliefs in music's elevated power.[80] In a letter to Tyndall on December 27, 1873, she wrote, "Nothing is more spiritual than music. . . . Listening to the noble compositions of Mozart, Beethoven, Handel . . . one loses all consciousness of matter to be absorbed in the purest and most spiritual of Delights."[81] Concerns about acoustical science also filtered into popular literature. The Cornish writer Nicholas Mitchell's poem "The Mystery of Music" (1869) urged readers, "Call music not mere vibrations / pulsing, trembling, floating by / Just to raise pleased, brief sensations."[82]

Acoustics also drew ire for its ties to the broader fin-de-siècle intellectual movement of physiological aesthetics (also called *Kunstphysiologie* or *esthétique scientifique*).[83] Though often forgotten today, physiological aesthetics held, according to Robert Michael Brain, "enormous cultural prestige" in its time, involving "the most high-profile, numerous, and highly funded network of laboratories of any scientific discipline."[84] Proponents of physiological aesthetics argued that human responses to art were fundamentally rooted in the muscles and nerves. The British psychologist James Sully, for instance, contended that physiological experiments were essential to providing the "precision" necessary "in our subjective estimate of sensation."[85] While Sully represented what Nicholas Dames calls a "soft" physiologist—one who resisted fully monist and mechanical views and maintained the role of the mind and spirit in aesthetic perception—he nonetheless believed in the importance of "scientific stud[ies] of sensation."[86] Similarly, in his landmark book *Physiological Aesthetics* (1877), the Canadian-born, English-educated scientist Grant Allen argued that the intensity of aesthetic feeling is directly related to the number of nerves excited by visual or auditory stimuli.

The fin-de-siècle writer Vernon Lee also intervened in the burgeoning field of physiological aesthetics. Though she took issue with what she saw as the field's almost-exclusive privileging of body over mind, she nonetheless believed in many of the main tenets of physiological aesthetics; as Carolyn Burdett writes, Lee could "never quite relinquish" the possibility that "science might eventually discover some hitherto obscure corporeal proof of how beauty affects us."[87] Lee read and heavily annotated Allen's book and conducted her own physiological experiments to trace and measure the sensations involved in aesthetic reception.[88] Along with her collaborator and partner Kit Anstruther-Thomson, Lee traveled to the museums of continental Europe and systematically recorded Anstruther-Thomson's bodily responses to paintings and sculptures, studying how looking at art altered the pace of her breath, the rhythm of her pulse, or the dilation of her eyes.[89] In the essay "Higher Harmonies," which appeared in her 1909 collection *Laurus Nobilis: Chapters on Art and Life*, Lee wrote, "Every day we are hearing of new discoveries connecting our æsthetic emotions with the structure of eye and ear, the movement of muscles, the functions of nerve centres, nay, even with the action of heart and lungs and viscera."[90]

Music was a major fascination for physiological aestheticians, and many of them engaged deeply with the work of Helmholtz, Tyndall, and other sound scientists.[91] In *Physiological Aesthetics*, which includes an entire chapter on "Hearing," Allen referenced the theories of Helmholtz and other sound scientists extensively in his discussions of the "pleasurableness of Musical Tones" and the "sympathetic vibrations aroused in Corti's organs."[92] Sully, who studied with Helmholtz in Berlin and attended many of his lectures (praising him as "the physiologist who has reached by far the most brilliant results in . . . objective analysis of sensation"), believed that "physical and physiological teaching" would offer better understandings of the "basis of musical pleasure."[93]

The ties between physiological aesthetics and music persisted for decades. In 1932, Lee published her landmark book *Music and Its Lovers: An Empirical Study of Emotional and Imaginative Responses to Music*, the outcome of a decades-long qualitative study of musical sensation that traced the various "bodily appeal[s]" of music for over 120 "music lovers," many of whom experienced "ecstatic and even orgiastic" responses to music.[94] As Lee wrote of one of her research subjects, "Cecilia," "If [music] goes on long her hands grow cold, she gets pins and needles and becomes breathless."[95] Another subject, Franz, described Wagner's music as inducing "tingling all over the body, beginning in the trunk and running down the legs and taking thirty seconds to spread through me."[96]

The ideas of physiological aesthetics—and its implications for music— were deeply polarizing. Many nineteenth-century intellectuals worried about attempts to "scientize" art and feared the "invasion" of aesthetics by natural science.[97] John Ruskin, for instance, believed that the study of art should have little to do with theories of "mere bodily gratification."[98] Physiological aesthetics threatened not only art's higher, metaphysical status but also notions of human autonomy and achievement. If music's effects could be explained by the process of sympathetic vibration and the reactions of the central nervous system, then what role was there for the genius composer, the virtuosic performer, or the sophisticated listener? In *On the Musically Beautiful* (1854), the Austrian music critic Eduard Hanslick expressed his skepticism about "an 'exact' science of music after the model of chemistry or physiology"; he worried that a composer would be nothing more than a "chemist who well understands how to mix all the elements of musical-sensual stimulation to produce a stupefying 'pleasure gas.' "[99] Similarly, the music critic François-Joseph Fétis argued that Helmholtz's theories would "annul the delicate sensations of the artistic ear for the benefit of essentially brutal calculations."[100]

Equally controversial was the idea that aesthetic experience could be automatic or preconscious, occurring without active intellectual engagement.[101] In *The Beautiful* (1913), for instance, Lee wrote that some (though not all) aesthetic pleasures were based on "bodily and mental reflexes in which our conscious activity, our voluntary attention, play no part."[102] For many thinkers, ideas of musical response as automatic and reflexive threatened the notion of the autonomous, thinking, and primarily *human* subject, reducing humans to almost animal-like beings governed by mere biological factors. If humans were all simply "automata," as T. H. Huxley controversially suggested, was there no room for human agency?[103] The German writer Max Nordau believed it was dangerous to understand sensual perception as something separate from conscious attention: "Untended and unrestrained by attention, the brain activity of the degenerate and hysterical is capricious and without aim or purpose."[104] He argued that the emphasis on humans' automatic bodily responses to art (including music) degraded humans to primitive creatures and brought them down "from the height of human perfection to the low level of a mollusc."[105]

This emphasis on musical automaticity was also controversial for its suggestion that musical response was universal—available to all listeners in equal measure. The British music critic Ernest Newman wrote in 1928, "I am quite willing to learn something new about Beethoven, but I have no time to waste in reading how Beethoven affects Tom, Dick, or Harry."[106] Democratic understandings of the universality of musical response aroused anxiety among physiologists and acoustical scientists as well. If even animals could respond in "extraordinary" ways to music, as the zoologist James Fennell wrote in 1841, then was musical response something that everyone could access, regardless of intellect, education, class, or race?[107] For some, the answer was yes. Havelock Ellis suggested that notions of musical response as preconscious or reflexive helped explain why even "an unmusical subject responds physiologically, with much precision, to musical intervals he is unable to recognize."[108] For Gurney, too, musical response was "universal, sensuous, and passive," available to everybody, even those who lack "musical ear[s]"; music, he believed, "speaks intelligibly and truly to the masses."[109] Elsewhere, Gurney suggested that people of the lower classes—those with "coarse and uncultivated natures," such as "East End roughs"—might even possess broader capacities for music appreciation: "Coarsely organized human beings seem comparatively little behind their more refined fellows in detecting the superiority of tone when they hear it."[110]

Not all acousticians were as democratically minded as Gurney, however. After all, acoustical science was largely indebted to an upper-class, Western musical sphere that did not take into account the tonal properties of Eastern or folk music. Several acoustical scientists went to great lengths to situate their theories within raced and classed hierarchies. Both Spencer and Helmholtz, for instance, privileged Western music as that most "naturally" suited to the human ear, insisting that "savage" music was *too* bodily, too dominated by base sensuality and devoid of skill or intellect.[111] Many acousticians believed that while musical sensation was available to everyone, the sophistication of musical response varied based on race, class, and species.

While some musicians, music critics, and even scientists were wary of acoustical science's perceived threats to spirituality, agency, and autonomy and of the democratic implications of automatic and reflexive musical response, many maintained that music's ties to physics and physiology represented a helpful way to understand, explain, and appreciate its powers and pleasures. Though, as mentioned earlier, Hanslick was ambivalent about physical theories of music, even he acknowledged that music should be investigated in part on "its corporeal side."[112] Similarly, Havelock Ellis insisted that understanding music's ties to "neuro-muscular tissue" could help humans better conceptualize the pleasures it diffused throughout the entire body: "Pleasure is a condition of slight and diffused stimulation, in which the heart and breathing are faintly excited, the neuro-muscular system receives additional tone, the viscera gently stirred, the skin activity increased; and certain combinations of musical notes and intervals act as a physiological stimulus in producing these effects."[113] Even nonmusical periodicals published defenses of acoustical science. In 1885, the *Weston-super-Mare Gazette, and General Advertiser* published a summary of a lecture given by a scientist named C. E. Frank, who argued that though materialistic understandings of music might seem "somewhat paltry and degrading" to the art, they had the potential to "elevat[e]" music: "Mechanical effects might prove of a higher character than was sometimes attributed to them."[114]

The mid- to late nineteenth-century intellectual world thus reverberated with heated conversations about what music—and art more broadly—is and does, and how humans and animals experience it through their senses. These discussions permeated almost every field of intellectual study, shaping how musicians, teachers, students, doctors, and writers thought about their work and the world around them.

Acoustical Science in Victorian Culture

In 1885, the Conservative monthly magazine the *National Review* published a fictional account of a lecture by an unnamed professor of physiology. In the story, the professor's opening provocation—"There is no such thing as music"—incites gasps and protestations from the audience members, who claim that they *know* music exists; they have heard it and have felt its power in the depths of their soul.[115] The "eminent man of science" goes on to explain what he means: What humans tend to think of as "music"—an abstract ideal, a pure essence that transcends the physical world—is merely a "superstitious notion," an "antiquated theory," and a "gratuitous hypothesis."[116] What music *is*, the professor argues, is a "sensation" resulting from vibrations that travel through the air, penetrate the nerve fibers in the ear, and affect the "grey nuclei" and "ganglionic nerve-cells" of the brain.[117] He explains that sound produces motor reactions in the human body that are both conscious (like dancing) and unconscious (like foot tapping).

As this anecdote reveals, by 1885, the acoustical ideas of Helmholtz, Tyndall, and their contemporaries—as well as their famously entertaining public lectures—were so ingrained in the Victorian imagination as to be fictionalized in a political newspaper. Music physics and physiology did not just circulate throughout the "British scientific elite" but also enjoyed a "robust . . . market" in Victorian popular culture.[118] Helmholtz's popularity in England is evident in the many public writings about him, particularly after his death in 1894. The satirical magazine *Punch*, for instance, memorialized Helmholtz by praising him as "Science's pride, and glory of thy land."[119]

Although, as Steege writes, acoustical science's seemingly "compulsive insistence on attention to acoustic minutiae" seemed a far cry from "'respectable' . . . upper-middle-class concert life," it nonetheless filtered into virtually every aspect of the nineteenth-century musical world, from music periodicals to performance manuals, conservatory curricula, and popular instruction materials.[120] While some musicians were threatened by the encroachment of scientific discourse into the musical world—as the English musician Swinderton Heap wrote in the *Musical Herald* in October 1893, "A physiologist cannot make a singer; that requires a musician"—many musicians embraced acoustical science as a way to enhance their own performances.[121] Singers were some of the most fervent adopters of physiological approaches. The voice teacher Emil Behnke and Dr. Lennox Browne, founder of the Royal National Throat Nose and Ear Hospital, co-published two physiological studies of the voice: *Voice Song and Speech* (1883) and *The Child's Voice* (1885). Behnke and his wife Kate also developed speech and breathing exercises and gave workshops throughout England.[122] The German mezzo-soprano Mathilde Marchesi was particularly famous for incorporating physiological study into her vocal training. Marchesi's teacher, the Spanish singer Manuel Garcia, was a scholar of vocal physiology; he wrote several books on vocal anatomy and invented the laryngoscope, a device that used a mirror to show the inner workings of a singer's throat. As Marchesi wrote in her autobiography, Garcia's physiological training was a "revelation" to her and the "foundation of [her] own future career."[123] Writing about Marchesi in 1898, the music critic Edward Baughan concluded, "The perfect teacher must have a thorough knowledge of physiology and acoustics, enabling him to fix the limits of each individual register and bind them together."[124] An understanding of acoustical science was now a crucial tool for successful musical performance.

While it was perhaps inevitable that physiological acoustics would eventually enter the realm of vocal music—after all, the body *is* the singer's instrument—instrumentalists were just as eager to embrace this new science in their performance practices. The nineteenth century witnessed the invention of several new devices geared toward the "physiology of pianoforte playing," as one description read.[125] Apparatuses such as the chiroplast, the dactylergon, and the technicon promised to help pianists strengthen their hand muscles and develop dexterity in their fingers.[126] Muscular development was essential to musical performance; the body could be adapted and trained to produce quality sound.

Physiological acoustics soon became a central component of Victorian music curricula. Book such as Sedley Taylor's *Sound and Music* (1873), John Broadhouse's *The Student's Helmholtz* (1881), and T. F. Harris's *Hand Book of Acoustics for the Use of Musical Students* (1887)—all of which addressed the theories of Helmholtz and Tyndall—were distributed to students at the Royal College of Music, Oxford, and Cambridge.[127] Many universities, including Cambridge, the University of London, and Trinity College, required acoustical study as part of their examination proceedings for musical degrees.[128] An 1891 advertisement in the *Musical News* for the degree of "Doctor of Music" at Oxford even specified that students could focus on "Helmholtz's Sensations of Sound" or works by Taylor.[129] Music physiology also made its way into more amateur music instruction. Many of Novello's Music Primers, educational manuals for young or beginning musicians that were marketed "at a price so low to render them attainable by all," contained sections on acoustics and physiology.[130] In 1878, Novello published a primer called *The Scientific Basis of Music* by W. H. Stone, a lecturer at St. Thomas's Hospital, which contained chapters on wave theory, harmonics, and overtones.[131]

Acoustical science influenced the field of medicine as well as music, with some doctors incorporating acoustical discoveries into their treatments. The mid- to late Victorian period witnessed what music historians refer to as the early "music therapy" movement, during which physicians at a range of institutions—such as the Cure and Nursing Home in Germany, the Earlwood Asylum in England, and the Perkins School of the Blind in Boston—began to use music to treat patients with physical and mental illnesses.[132] An anonymously published 1881 article in the *Musical Times and Singing-Class Circular*, titled "Music as a Relief to Pain," refers to a French surgeon named Mr. Vigouroux who "administer[ed]" sound waves as a cure by holding tuning forks and sounding boards near affected parts of the body.[133] Similarly, an October 1880 article in the *Musical Times* by Dr. Henry C. Lunn celebrates music's "sensible influence on the circulation of the blood" for ailing patients.[134] In 1892, the physician J. Ewing Hunter wrote in the *British Medical Journal* of an experiment he conducted at his hospital in which seven out of ten patients exposed to music experienced fever reduction.[135]

Perhaps most prominent among these late nineteenth-century "music therapy" efforts was the Guild of St. Cecilia, founded in 1891 by Frederick Kill Harford, an Associate at the Royal College of Music.[136] The guild brought musicians into a variety of London hospitals and mental institutions to alleviate patients' suffering. The *British Medical Journal* and other periodicals widely publicized the work of the guild, and medical figures like Florence Nightingale and Sir Richard Quain (Queen Victoria's physician) praised the organization.[137] By the end of the century, the burgeoning "music therapy" movement had grown so prominent as to inspire satirical treatment from the *Musical Times*:

> Doubtless we shall come in time to have electric call-boxes provided with a special signal for summoning the medical musician, so that within a very few minutes of the first symptoms of influenza, or mumps making their appearance we shall be able to nip the ravages of these maladies in the bud by application of the proper musical remedies. . . . The supporters of the new departure can point in triumph to the etymology of the word Music (=Muse sick).[138]

Victorian educators also seized on new understandings of the human body's responsiveness to music, particularly in their implementation of new disciplinary systems. Musical drills—thought to help train students' bodies and minds—became integral parts of many educational programs. Late nineteenth-century physical education manuals such as R. H. McCartney's *Gill's Physical Exercises or Dumb Bell Drill with Musical Accompaniments* (1881) and *Barbell or Wand Exercises* (n.d.), Mrs. L. Ormiston Chant's *Golden Boat Action Songs*

46 | Sounding Bodies

(1890), Flora T. Parsons's *Callisthenic Songs Illustrated* (1869), and F. Leslie Jones's *Songs and Choral Marches for School Use* (1875) put new ideas about the kinesthetics of music to practical use. Exercise 14 in *Barbell or Wand Exercises*, for example, directs students to complete "one movement to each bar of music" (see fig. 1.8).[139] Similarly, Parsons advocated "musical exercises" as "pleasant and profitable" pursuits for students: "Let them rise upon their feet, shake off the shackles of drowsiness, clap their little hands, and join in song, and instantly, thoughts are invigorated and quickened, and all are prepared for efficient work."[140] As mentioned in the introduction, however, these "educational" uses of music often had troubling classist, racist, and colonialist implications. Steege, Charles McGuire, and Bennett Zon have argued that music education systems such as John Curwen's Tonic Sol-Fa exercises, in which vocal exercises were paired with physical ones (marching in place, raising and lowering the arms, turning the head), often served as forms of paternalistic control for children, workers, and colonized peoples, forcing them to become, Steege writes, "an army of Foucault's proverbial 'docile bodies.'"[141] Acoustical science had circulated so widely throughout Victorian culture that it could be mobilized for almost any ideological agenda.

Figure 1.8. Exercises, in Flora T. Parsons, *Callisthenic Songs Illustrated* (New York: Ivison, Blakeman, Taylor & Co., 1869), 3. *Source:* Archive.org. Public domain.

By the end of the nineteenth century, acoustical science had seeped into almost every aspect of Victorian life. It is not surprising, then, that new advancements in sound science also made their way to Victorian writers—and readers. The fictional texts I discuss in subsequent chapters absorbed these new acoustical ideas in order to delve inside their characters' bodies and depict their most intense corporeal sensations. The ideas of Helmholtz, Tyndall, and many others enabled Victorian authors to render their characters human versions of Tyndall's magic wands—shaking, shivering, dancing bodies overcome by the material powers of sound.

Part Two

Genders

Chapter Two

Bare Arms and Quivering Nerves
The "Lady Violinist" Novels of
Mary Augusta Ward and M. E. Francis

On the surface, Rose Leyburn is the ideal musical heroine. One of two female protagonists in Mary Augusta Ward's 1888 novel *Robert Elsmere*, Rose performs at the homes of British socialites and charms her family members and suitors with her music. Her musicality reflects her education, refinement, and social status as she displays the "feminine accomplishments" expected of middle- and upper-class Victorian women.[1]

Yet Rose deviates from this ideal in three crucial ways: her choice of instrument, her pursuit of a professional career, and her intense bodily involvement in her music. First, Rose is patently *not* the "heroine at the piano" beloved in Victorian culture.[2] Rather, she plays the violin, an instrument from which women were all but "ban[ned]" for much of the nineteenth century due to its grotesque construction (animal-gut strings), its reputation as the "devil's instrument," and its involvement of awkward and unattractive movements of the face and body.[3] Moreover, Rose far surpasses the amateur status expected of nineteenth-century female musicians. She practices for hours a day; plays an infamously difficult repertoire; travels to Manchester, London, and Berlin to train with renowned violin masters; and performs in public with professional ensembles. Finally, and most subversively, Rose plays her violin with fierce bodily intensity. She throws herself into her music, vigorously sawing at the strings and setting her listeners' "every nerve . . . vibrating."[4]

This chapter explores *Robert Elsmere* alongside another late-Victorian "lady violinist" novel: *The Duenna of a Genius* (1898) by M. E. Francis (née Mary Elizabeth Blundell).[5] Francis's heroine Valérie is also a virtuosic violinist who pursues a professional performance career. Like Ward, Francis presents vivid, extensive, and recurring illustrations of the physiological powers of her heroine's violin playing. Valérie possesses instinctive musical virtuosity and harnesses the vibratory powers of sound to induce pleasurable sensations in her own body and in the bodies of her listeners.

Both *Robert Elsmere* and *The Duenna of a Genius* mobilize the language of acoustical science to depict and defend the controversial practice of female violin playing in the Victorian era. Rose and Valérie wield their strong arm muscles as they draw their bows across their violin strings, seize control over performance spaces as they send their audiences into raptures, and grant themselves visceral pleasures as they play and sway. In such moments, *Robert Elsmere* and *Duenna* intervene in the gender politics of the Victorian musical world, which were becoming increasingly vexed in the latter half of the century as more and more women began to make music in quasi-professional or professional settings. Discussed in further detail below, Victorian music critics often denigrated female violin playing *because* of its involvement of the body, deeming it either unacceptable (involving inappropriate levels of bodily display) or physically impossible (necessitating an amount of muscular strength or agility that women allegedly did not possess).[6] Yet the science of physiological acoustics—in particular, new understandings of musical talent as an innate bodily trait and of music appreciation as an automatic and universal physical response—empowered Ward and Francis to cast female violin playing not as a deviant or impracticable pursuit but as one safely supported by contemporary acoustical research.

By incorporating acoustical language of nervous and muscular sensation, Ward and Francis also present overt and extensive depictions of female erotic experience—erotic in Lorde's sense of an "internal sense of satisfaction," a "life force," or "creative energy empowered."[7] Rose and Valérie find their greatest sources of sustenance not in their domestic activities, nor (it is implied) in their sexual relationships with their husbands, but in their violin playing. Acoustical science helps Ward and Francis illuminate how, for the rare women who are able to play, music making can be the most erotic experience in the world.

Below, I trace the rise of female violin playing in late-Victorian England and the virulent responses that these "lady violinists" incited. I then show how Ward and Francis drew upon contemporary acoustical science to intervene in these discourses, articulating a musical gender politics that locates women's musical abilities as rooted in their bodies. I close by showing how *Robert Elsmere* and *Duenna* anticipate the work of twenty-first-century feminist and queer science theorists who argue for the utility of biological and physiological thinking for feminist and queer politics. Physiological science is not always antagonistic to radical ideas about gender but can in fact be crucial for their articulation.

The "Lady Violinist" in Victorian Literature

In their appeals to the language of acoustics to celebrate their heroines' violin playing, *Robert Elsmere* and *Duenna* deviate from most other Victorian literary representations of female musicianship. Works of Victorian fiction most often depict women as passive music *listeners*, possessing the appropriate taste and training necessary to appreciate good music and to impress suitors with their knowledge. When women do create music on their own, they usually appear as drawing-room singers or "heroines at the piano," an instrument that, Gillett writes, "accorded well with female modesty" and required "no awkward motions or altered facial distortions."[8] While women's amateurish relationships to music at times become a source of frustration for characters (and authors), as when *Daniel Deronda*'s Herr Klesmer expresses his "lofty criticism" of Gwendolen Harleth's lack of high-level musical ability (a moment that reflects Eliot's own distaste for dilettantish artistic pursuits), this amateurism more often than not exemplifies women's moral virtue.[9]

Fictional women musicians who do depart from the "heroine at the piano" trope often appear as "sirens" or "divas." Weliver and Delia da Sousa Correa argue that "siren" narratives, often found in works of sensation fiction such as Mary Elizabeth Braddon's *Lady Audley's Secret* (1862), intertwine musical storylines with plots of deception, infidelity, and seduction, in which musical ability becomes a stand-in for women's questionable morality.[10] In *Lady Audley's Secret*, Lady Audley's music serves as a crucial part of her dangerous "masquerade."[11] "Divas," on the other hand, often ruthlessly pursue their artistry at the expense of their other feminine duties (marriage and motherhood) and their moral virtues. Many of these "affected females," as Emily Auerbach calls them, ultimately fall prey to their own hubris; Eliot's Alcharisi and Armgart, for instance, are both hyper-ambitious virtuosi who lose their voices at the height of their careers, destined either to die bitterly, lamenting the loss of their artistic skill (Alcharisi), or to be forced to transform their passion for performing into something more outwardly oriented or ethically inflected, like teaching (Armgart).[12]

Novels about female violinists did not appear until late in the nineteenth century, when women's violin playing became marginally less taboo due to the growing fame of performers like Wilma Norman-Neruda (Lady Hallé) and the gradual opening of music conservatories to female string players. When they did appear, most "lady violinist" novels rehearsed the siren and diva stereotypes found in earlier texts. George Gissing's *The Whirlpool* (1897), Cecily Sidgwick's *A Splendid Cousin* (1892), Joseph Conrad's *Victory* (1915), Lady Mabel Howard's "Forgotten Chords" (1897), and Albert Morris Bagby's *Miss Traumerei* (1895), for instance, imagine female violinists whose musicality renders them sensual, narcissistic, demonic, or dangerous.[13] Other narratives fantasize about the thwarting of a female violinist's career before it has even begun. In Mildred Finlay's "The Stradivarius" (1897), for instance, a mother is so afraid of her daughter's instrumental pursuits that she burns her violin.

Some Victorian texts do offer slightly more sympathetic treatments of female violinists. The protagonists of Jessie Fothergill's *The First Violin* (1877), Walter Besant's *Armorel of Lyonesse* (1906), Louise Mack's *The Music-Makers* (1914), Elizabeth Godfrey's *Cornish Diamonds* (1895), and Daisy Rhodes Campbell's *The Violin Lady* (1916) flirt with the possibility of musical careers for their violin-playing heroines. Yet the female musicians in each of these stories are ultimately distracted by romantic pursuits or decide to give up their music when they marry. In *The Violin Lady*, for instance, Virginia decides that her husband Alan is "the only man in the world for me, the only one I could ever have put before my violin, 'once the sole master of my heart.' "[14]

Robert Elsmere and *Duenna* present striking alternatives to these narratives. Both Rose and Valérie stave off the traditional demands of marriage and motherhood for as long as they can in order to pursue their music, and though both women do ultimately marry, neither abandons her music after she weds. Valérie in particular enjoys a wildly successful, lifelong musical career. Most importantly, though, *Robert Elsmere* and *Duenna* tinge their scenes of female violin playing with the acoustical language of sound waves, sympathetic vibration, and nervous response. The "lady violinist" novel could thus be a site for subversive gender politics rather than a vehicle for the misogynist anxieties that, as the next section shows, dominated the nineteenth-century musical world.

"Beyond Her Muscular Power": Female Violinists in Victorian England

The fin-de-siècle backlash against female violinists was spurred by their growing visibility in Victorian society. As performers like Wilma Norman-Neruda, Marie Hall, Emily Shinner, Teresina Tua, and Trida Schytte earned acclaim across England and Europe, more and more women began to learn the violin (see fig. 2.1).[15] The American violinist Edith Lynwood Winn wrote in the *Musical World* in 1901, "Women have won renown as virtuosos, and this alone ought to spur on the young girl violinist. . . . The careers of these women are full of inspiration."[16]

The growing interest in female violin playing was also stimulated by the gradual opening of music conservatories to women toward the end of the century. During this time, the Royal Academy of Music, Royal College of Music, and Guildhall School of Music—eager to fill their tuition coffers—all opened slots to women, who soon made up the "immense majority" of the students at these institutions.[17] By the 1880s, it was no

Figure 2.1. Painting by James Tissot, *Hush!*, c. 1874, oil on canvas, 73.7 × 112.2 cm. *Source:* Wikimedia Commons. Public domain.

longer an "uncommon sight in our streets to see a girl carrying her fiddle in its black case," remarked the Victorian-era violinist and patroness of the arts Lady Blanche Lindsay.[18] And by the end of the century, according to the violinist, music critic, and poet Marion Scott, "the 'lady violinist' was an accepted, though still rather striking fact."[19]

Despite these advancements, very few women were actually able to forge the kinds of virtuosic solo careers enjoyed by figures like Norman-Neruda, and almost none made their way into mainstream symphony orchestras. A November 1906 article in the *Musical Times* by a writer named "F. G. E." reflects just how exceptional a case like Norman-Neruda's was: "Violin playing by ladies made slow progress in England, even after the wonderful achievements of Wilhelmina Neruda (Lady Hallé) gave it such a splendid impetus."[20] Though dozens of "ladies' orchestras" formed to offer performance opportunities to this new glut of skilled but unemployed female instrumentalists, advocates like the violinist Ada Molteno believed that women's musical achievements would be of "little use" until they were systemically admitted to mainstream professional orchestras.[21] While Henry Wood, director of the Queen's Hall Orchestra, made the radical move to hire six female string players in 1913 (due to a shortage of male players during World War I), it was only after World War II that women were really able to, as musicologist Lucy Green writes, "flood the major orchestras."[22] According to the London Symphony Orchestra historian Richard Morrison, the LSO remained an "enclave of lads" as late as the 1970s.[23]

The archive of late nineteenth-century music criticism reveals several justifications for this systemic exclusion of female violinists from solo and orchestral careers. Some blamed the demands of marriage and motherhood for overtaking women's musical careers, as occurs in Campbell's *The Violin Lady*. Winn wrote, "Very few of our women violinists pursue their professional life after marriage. The profession, therefore, does not become overcrowded, for women, even violinists, will meet the artful Cupid now and then."[24] Other music critics warned that solo or orchestral careers would hamper women's moral sensibilities. The pianist W. Tyacke George, for instance, was concerned that "a rough apprenticeship at theatres and music halls would affect an educated, refined, and sensitive girl physically, and probably morally," especially with the "unguarded language which the theatre employees make such liberal use of within earshot of the band room."[25] Exposure to this kind of Victorian-era "locker-room talk," some worried, could result in dire consequences for women's femininity. As Lady Lindsay wrote, "I have . . . known girls of whom it was darkly hinted that they played the violin, as it might be said that they smoked big cigars, or enjoyed the sport of rat-catching."[26] Further amplifying this sense of danger was the common nineteenth-century perception of the violin as the "devil's instrument"—a centuries-old association drawn from folk tales and ancient legends that linked violin playing to satanic rituals and deals with the devil.[27] Cigar smoking and rat catching were thus among the least abhorrent outcomes of professional musicianship.

The fiercest objections to female violin playing, however, concerned its ties to the body. Fears about public musical performance for women harkened back to centuries-old associations between the theatre and prostitution—of stage performance as a form of bodily exhibition.[28] Though the East End music-hall stage might have tolerated titillating spectacles by (often lower-class) women, the logic went, the realm of high-art classical music—an important site for the British middle and upper classes to build up their cultural capital—could not accommodate such provocative displays.[29] The female pianist could delicately tickle the keys with the tips of her fingers, and the female singer's vocal apparatus was hidden from view, but the violinist needed to raise her arm, assume an active posture, rest the violin against her chin, and move vigorously, with several parts of her (dynamic, moving) body on full display. Music critics worried that showcasing this active body in public could subject women to the leering gazes of male spectators.[30] George Du Maurier's *Punch* cartoon "The Fair Sex-Tett (Accomplishments of the Rising Female Generation)" (1875) not only satirizes female musicianship as ridiculous (note the absurdly large brass instrument on the right side of the frame) but also highlights the voyeurism of the men in the audience (see fig. 2.2).[31]

The cello, as one might imagine, was particularly objectionable, as it required performers to spread their legs in order to hold the instrument—a gesture that obviously evoked a sexual pose. In his manual *The Technics of Violoncello Playing* (1898), the music critic Edmund S. J. van der Straeten wrote that, for many years, the "graceful" way for a lady to hold the cello was to play sidesaddle—to "turn both legs to the left, bending the right knee and placing it under the left one" (see fig. 2.3).[32] In fact, a major motivation for the midcentury invention of the endpin, a rod that attaches to the bottom of the cello and secures it to the ground, was that it would allow women to play the instrument without spreading their legs.[33]

While string playing could make women too sensual, it could also make them too ugly, since it required the kinds of "awkward motions or altered facial distortions" that instruments such as the piano did not neces-

Figure 2.2. Cartoon by George Du Maurier, "The Fair Sex-Tett," c. 1875, pen and ink, from *Punch* 68 (April 3, 1875): 150. *Source:* Archive.org. Public domain.

Figure 2.3. Painting by Thomas Wilmer Dewing, *Lady with Cello*, 1920 or before, oil on canvas, 20⅛ × 16¹/₁₆ in. *Source:* Bequest of Annie Swan Coburn (Mrs. Lewis Larned Coburn), from Smith College Museum of Art, Northampton, MA, SC 1934.3.4. Used with permission.

sitate.[34] The cartoon "Joachim's Rival," published in 1880 in the *Musical World*, caricatures Norman-Neruda as an unattractive "creatress" with a large, cartoonish head and nose—a "rival" of the "darling" Hungarian violinist Joseph Joachim (see fig. 2.4).[35] Remarks on the unattractiveness of female violinists also appeared in real-life reviews. In 1834, the *Athenaeum* wrote that though the German violinist Elise Filipowicz possessed "sufficient skill and feeling to give our ears great pleasure . . . our eyes told us that the instrument is not one for ladies to attempt."[36] Similarly, in *Musical Memoirs* (1830), the oboist William T. Parke wrote that the French violinist Madame Louisa Gautherot "should display her talent in a situation where there is only just light enough to make 'darkness visible.'"[37]

To many, female violin playing was not only inappropriate or unattractive but also biologically impossible. On the heels of Darwin's *Descent of Man* (1871), sexist ideas about women's inferiority were "'reissued' in powerful scientific packaging," Gillett writes, and some music critics and scientists insisted that women were physically incapable of performing at a

Figure 2.4. Cartoon, "Joachim's Rival," c. 1880, from *Musical World* 10, no. 58 (March 6, 1880): 153. *Source:* HathiTrust Digital Library. Public domain.

high level.³⁸ Many doubted that women possessed enough bodily strength to accommodate the heft of an instrument or the amount of physical vigor required to play it. In a February 1894 piece in the *Orchestral Association Gazette*, the music critic Wallace Sutcliffe insisted that women were "constitutionally" incapable of the "heavy, arduous labour" required by musical performance: "Lack of physical power has always and will always hamper women in any arduous work. According to physiologists, there is one muscle entirely absent from the female arm. Thus, power of tone is always lacking. If we take our great instrumental virtuosi as an example, do we ever find the same strength, power, and breadth of tone in the female exponents as in the male?"³⁹ Similarly, W. C. Honeyman, a violin professor, suggested that female violinists were inhibited by their lack of muscular strength, which rendered them unable to produce anything but weak, thin tones; if a female violinist "force[d] the tone beyond her muscular power," she would emit "harsh noise."⁴⁰ Honeyman also argued that women's hands were smaller than those of men, rendering them unable to play wider intervals (such as fifths and octaves) and larger chords.⁴¹

Discussions of the female violinist's limited strength and feeble tone persisted well into the twentieth century. As late as 1928, when the Society of Women Musicians (SWM) wrote to the BBC's director general, Sir John Reight, to ask why the new National Orchestra of the BBC had no female members, director of music Percy Pitt replied that "men were preferred because their tone was bigger and stronger."⁴² Pitt noted that Sir Henry Wood's orchestra (the ensemble that admitted women during World War I) was "the weakest in London."⁴³

Others warned that violin playing was too physically taxing, even dangerously so, for women. In the same article in which Dr. Lunn praised music's beneficial effects on patients, he noted its "exhausting effect" on young girls, which he believed would "shortly engage the attention of the medical profession"; the "months of hard study" that violin virtuosity required were better left to the "more vigorous brains of male students."⁴⁴ Similarly, in a fascinating weaponization of acoustical science *against* female performers, J. Herbert Dixon wrote in the *Medical Magazine* in 1900 that women pianists risked enduring "the baneful influence of the continual vibrations on the organ of Corti and so on the brain" and thus experiencing "headaches, neuralgia, nervous twitchings, hysteria, melancholia, madness."⁴⁵ Tyacke George—the same man who worried about the moral corruption awaiting women backstage—claimed that "protracted and wearying" orchestra rehearsals would tax the female body: "The duties of an orchestral player are ever arduous, sometimes exacting to a degree; and if fatigue and occasional exhaustion are now experienced by those who perform these duties, how much greater would be the sufferings of the feebler sex, whose constitutions are naturally delicate and comparatively weak? I believe that a positive breakdown would be in many cases the inevitable result."[46]

Several late-Victorian activists and music critics protested these codes about female musicianship. In 1900, the (aptly named) musicologist Florence G. Fidler wrote a letter to the *Musical News* insisting that women would make more capable orchestra members than men: "Women are tired of living under glass-cases. . . . Experience has taught me that . . . as a general rule, a woman is more punctual, more attentive to the conductor, and in every way discharges the work she is paid to do more conscientiously than does a man."[47] Joseph Barnaby, a composer and conductor, though skeptical that women could harness the full "nerve strength, and endurance [to] . . . play through a long and trying programme," nonetheless argued that "the experiment might be tried, and would be a good thing in that it opened another career to women."[48]

Others more fervently insisted that female violinists were just as physically strong as male players. Molteno called Sutcliffe's "'missing muscle' argument" "distinctly humorous," and instructors like Lady Lindsay urged their female students not to internalize erroneous ideas about their lack of strength and instead to prove detractors wrong by developing their musical muscles: "Practise regularly, even though you are disinclined; unless you are *really* ill, a little weariness or fatigue soon goes off, and after playing for ten minutes you will probably feel fresher than before you began."[49] Some reviewers specifically highlighted the immense strength that famous female violin players displayed. One 1888 reviewer for the *Musical Standard* wrote that the Swedish violinist Anna Lang used her bow "with rare power."[50] Similarly, Camilla Urso's biographer Charles Barnard emphasized that Urso worked from childhood to achieve the physical strength required to play her instrument:

> She must first learn to sustain the weight of the violin, and accustom her arm to its shape. . . . Then, that right wrist! How it did ache with the long, slow motions with the bow. And her limbs grew stiff with standing in one position till they fairly ached. . . . As the weeks grew to months, her fingers and arms gained in power. . . . Strength, power, and purity of tone were the things worth trying to reach. She would have no feeble, short strokes, but the wide, bold movements of a master hand.[51]

Urso's displays of strength, according to Barnard, "demoralized" the "anti-Urso party . . . and they gracefully retired from the field."[52] *Robert Elsmere* and *The Duenna of a Genius* present literary versions of such defenses, drawing upon acoustical science to resist the notion that music was "beyond [women's] muscular power." Acoustical science provides Ward and Francis with an empirical basis to depict heroines who possess innate musical talent, perform with vigor, and excite their listeners' bodies with their music.

Brahms in the Blood: *Robert Elsmere*

On the whole, music plays a rather minor role in *Robert Elsmere*. Both Victorian and modern-day readers of Mary Augusta Ward's novel have focused on the crisis-of-faith plotline of its titular protagonist, an Oxford-educated clergyman who leaves the Anglican faith to pursue a "liberalized and humanized" religion.[53] Scholarly attention centers mostly on the novel's engagement with Christian moral thought and its exploration of the tensions between religion and marriage.[54]

As a result, critics have focused less on *Robert Elsmere*'s equally suggestive secondary narrative, that of Robert's violin-playing sister-in-law, Rose. Though Ward herself was passionate about music—having studied organ at Oxford's New College and founded a music program at the Passmore Edwards Settlement—she rarely mentions Rose in her later prefaces or other writings.[55] As Weliver points out, even the Victorian music critic H. R. Haweis "avoid[ed] any mention of the subplot" in his foreword to the novel.[56] While several critics have highlighted Rose's passionate spirit and her rejection of strictures of Victorian femininity, and others have lamented her ultimate marriage to the philanthropist Hugh Flaxman as evidence of her failed feminism, few have explored her musicianship at length.[57]

And yet, Rose's violin playing propels one of *Robert Elsmere*'s most crucial interventions: its defense of professional female musicianship. Ward herself was apprised of new developments in physiological science. References to anatomy and physiology ("tingling" nerves and "sympathetic" and "vibrating" "fibres") are peppered throughout *Robert Elsmere*—knowledge that perhaps came from the Ward family's frequent dining with physiologists and acousticians like T. H. Huxley, Lord Rayleigh, and R. H. M. Bosanquet.[58] Ward was thus well positioned to depict a performer who affects her listeners' muscles and nerves and gratifies her own body while playing.

From the beginning of the novel, Rose chafes against Victorian codes about female musicality. Ward explicitly distances Rose from the drawing-room heroines that dominated much Victorian fiction by associating her with a serious, high-cultural, and Continental musical sphere. When Rose's sister Catherine urges her to devote herself to more traditional domestic or religious activities or to pursue more traditional musical paths such as teaching, Rose indignantly replies, "When one can play the violin and can't teach, any more than a cockatoo, what's the good of wasting one's time in teaching?"[59] Unlike her amateur acquaintances—and her dilettantish counterparts elsewhere in Victorian fiction—Rose emerges as a diligent practitioner of her art. Rose practices for hours and rejects the kinds of drawing-room repertoire—mass-produced and widely available to Victorian middle-class households—that most women performed at the time.[60] Instead, she plays works by Robert Schumann, Johannes Brahms, Richard Wagner, and Louis Spohr, a Romantic-era composer known for writing solo violin works so technically difficult that even experienced performers rarely undertook them (Rose's playing of Spohr, however, is "nearly perfect").[61] When others denigrate her musical pursuits as unwomanly—the rector's wife Mrs. Seaton calls the violin an "unbecoming instrument for young women"—Rose weaponizes her sophisticated musicality against them.[62] After a particularly "shrill[]" performance by a local duet, Rose "wickedly" plays one of Schumann's *Fantasiestücke* to provide "the most severely classical contrast to the 'rubbish' played by the preceding performers."[63]

Though some Burwood townspeople call Rose a "village Norman-Neruda"—a moniker that acknowledges her talent but restricts her to a local sphere—her musical reach extends far beyond her provincial home.[64] Rose travels to Manchester, London, and Berlin, where she embeds herself within a cosmopolitan network of esteemed performers, whose names Ward drew from the real-life musical world. In London, Rose meets a "crowd of artistic friends" that includes the soprano Emma Albani and plays Brahms with a pupil of the pianist Anton Rubinstein.[65] In Berlin, Rose plays in a quartet with a German violist and a Hungarian cellist and even gets the renowned Hungarian violin instructor Joseph Joachim to teach her and give her "unusual attention."[66] Rose moves freely through artistic circles, finding them not dangerous or inhospitable but supportive and affirming.

Rose is not just a successful musician; her musicality is inextricable from her physical life. Ward makes clear that Rose possessed musical proclivities even as a young child. In one early scene, Rose physically cannot stand the shrill or out-of-time playing of her fellow townspeople; she murmurs that the players are "two bars ahead" and stamps her foot in response to their "rubbish" sounds.[67] Later, the narrator notes that Rose has "Wagner and Brahms in her young blood," a phrase that ties her musical artistry to the workings of her circulatory system.[68]

Rose experiences violin playing as a reflexive and instinctive process. In one scene, when she plays an "Andante and Scherzo" by Beethoven, she appears wholly consumed by her music, but in a way that is natural rather than wanton or depraved: "The art of it was wholly unconscious. The music was the mere natural voice of her inmost self."[69] This moment invokes Victorian-era scientific (and pseudoscientific) writing about musical ability as an innate quality. In his essay *Wer ist musikalisch? (Who Is Musical?)*, Theodor Billroth, a German surgeon and music aficionado (and friend of composer Johannes Brahms), drew on Helmholtz's acoustical theories to propose that musical abilities, such as the capacity to replicate melodies and follow rhythms, were intrinsic skills determined at birth.[70] The phrenologist (and now-infamous scientific racist) François Joseph Gall proposed that musical talent was determined by the "cerebral organization" and the size of the "organ of tones."[71] After meeting a young prodigy named Bianchi, Gall theorized that "the talent for music could be recognised by the form of the head."[72] Phrenological ideas of musical talent as innate persisted into the twentieth century. In "The Measurement of Musical Talent" (1915), the American psychologist Carl Seashore described musical ability as "inborn" and argued that musical skill could be measured and predicted; he even suggested that "laboratory specialists" could help identify musical talent in "those who have serious aspirations for a professional career in music."[73] In *Robert Elsmere*, Ward depicts Rose's music as involuntary, rooted in her body, and fundamental to her physical life—a set of intrinsic skills apart from the world of the intellect ("unconscious"). Rose's violin playing is neither unacceptable nor unnatural nor impossible; in fact, it is by playing Beethoven (rather than, say, child rearing or other activities commonly deemed "natural" to Victorian women) that Rose is best able to access what is most natural to her. The compulsion—and the ability—to play Beethoven is embedded in her body.

Rose's musical abilities grant her the power to affect the world around her. She takes up acoustical space in ways that are palpable to her and her listeners. In one scene, she "play[s] with all her soul, flooding the house with sound—now as soft and delicate as first love, now as full and grand as storm waves on an angry coast."[74] The language of "waves" straddles the figurative and the literal. While certainly a metaphor for her passionate performance, this language also invokes contemporary scientific discussions of sound's undulatory motion through the air, as when Tyndall's singing flames proved "powerful enough to shake the floor and seats, and the large audience that occupies the seats of this room."[75] Rose's music—loud and forceful—similarly reverberates through her physical surroundings; the laws of physics account for her ability to flood the house with sound. Rose's music takes on a concrete presence as she produces actual waves that penetrate the atmosphere and reverberate through space.

These waves of sound wash over Rose's listeners' bodies as well as her physical environment. When her future love interest Edward Langham first hears her play the works of Wagner, Brahms, and Rubinstein, he feels the music "rippl[e] over him in a warm, intoxicating stream of sound. . . . What magic and mastery in the girl's touch!"[76] Again, metaphorical and scientific language merge to depict Edward's experience as at once transcendent and deeply sensual. In the context of Victorian acoustics, Rose possesses the capacity to physically overwhelm Edward and to awaken his sensations, dulled by intellectual study. Her sound moves in waves—"ripples"—that enter the ear, strike the tympanic membrane, and affect the entire body, reaching Edward as a form of "touch."[77]

Rose's musical influence becomes even more acute when she again performs for Edward later in the novel. She "thr[ow]s herself" into an "Andante and Scherzo" of Beethoven "in a way which set every nerve in Langham vibrating!"[78] The passage highlights the patently corporeal nature of Rose's performance. She plays with physical abandon ("throwing" her whole body into the music), and her performance affects Edward on a visceral level ("vibrating" his "nerves")—details that recall contemporary acoustical discussions of the nervous and muscular energy required of performers and the vibratory effects of sonic stimuli. Edward's nerves, we might say, are vibrating sympathetically to Rose's music. The image of his vibrating nerves is comprehensive ("*every* nerve") as well as concrete. Edward marvels that, though "his life had been so long purely intellectual," Rose awakens in him a "sudden strain of passion . . . [that] seemed to unhinge him, to destroy his mental balance."[79] While Edward lives a life of the mind, Rose's music ignites his body.

Crucially, though, it is not just Rose's love interest who is physically affected by her music. In one scene, Rose performs for a group of London socialites, who expect her to play piano.[80] Instead, Rose retrieves her violin and plays "some Hungarian melodies put together by a younger rival of Brahms" that send the whole party into physical raptures:

They had not played twenty bars before the attention of everyone in the room was more or less seized. . . . 'Ah, but *excellent!*' said Lady Charlotte once, under her breath, at a pause; 'and what *entrain*—what beauty!' . . . Then the slow, passionate sweetness of the music swept her [Lady Charlotte] away with it, she being in her way a connoisseur, and she ceased to speculate. When the sounds ceased there was silence for a moment. Mrs. Darcy, who had a piano in her sitting-room whereon she strummed every morning with her tiny rheumatic fingers, and who had, as we know, strange little veins of sentiment running all about her, stared at Rose with open mouth.[81]

The reference to the attention "seized" within "twenty bars" casts the effects of Rose's music as immediate and inevitable.[82] The music takes Lady Charlotte out of a conscious mental realm ("she ceased to speculate") and propels her to mutter excitedly under her breath. Her mental awareness evaporates, leaving only reflexive, bodily action. Eventually, the music stuns the audience members into silence, leaving them with mouths agape. The image of Mrs. Darcy's "strange little veins of sentiment running all about her" nods to the effects of Rose's music on her nervous and circulatory systems. Rose's violin playing is thus not simply a conduit for her romantic courtship with Edward but a much stronger force that arouses a variety of listeners.

Readers again witness the universal—and patently physical—effects of Rose's music on her listeners in a later scene, when she and Edward play a duet for an audience of English and Continental musicians. Not only is this moment brimming with personifications and metaphors that capture the music's capacity to take up physical space—"How that Scherzo danced and coquetted, and how the Presto flew as though all the winds were behind it, chasing its mad eddies of notes through listening space!"—but the performance also invites an actual "wild storm of applause" and awakens "guttural sounds of delight" among audience members.[83] Music appreciation comes from deep within the body, and Rose is able to affect her listeners at the most "guttural" level.

Rose awakens corporeal sensations in her listeners, but she also grants *herself* experiences of strength, energy, and pleasure as she performs. These sensations appear not as dangerous or depraved but as healthy and natural, explained by physiological acoustics. Rose relishes the sense of invigoration that playing provides: "She stood with her lithe figure in its old-fashioned dress thrown out against the black coats of a group of gentlemen beyond, one slim arched foot advanced, the ends of the blue sash dangling, the hand and arm, beautifully formed but still wanting the roundness of womanhood, raised high for action, the lightly posed head thrown back with an air."[84] While in some ways this passage highlights Rose's docility and femininity—her "slim" foot, her dress with a sash, and her "beautiful" form—it also emphasizes the profound physicality of her playing. Rose assumes an active, commanding pose that involves several parts of her body (foot, arm, and head) and displays corporeal dynamism, raising her arm "high for action." Crucially, her vigorous bodily movements are seen not as unattractive but as "bewitching . . . overflowing with potentialities of future brilliance and empire."[85] Rose assumes a similarly sophisticated posture later in the novel, "standing a little in advance of the others, her head turned to one side, really in the natural attitude of violin-playing . . . the right arm and high-curved wrist managing the bow with a grace born of knowledge and fine training."[86]

Rose's violin-playing pose in these moments reflects her thorough musical training and her body's adaptation to the instrument. Lady Lindsay wrote in "How to Play the Violin," for instance, "It is *very* difficult to bow well; to hold the bow aright, lightly . . . to keep the thumb steady, and the four fingers straight (not curved outwards), the tips resting firmly on the bow."[87] As a result, many nineteenth-century violin manuals focused on teaching players how to adapt and manipulate their bodies to achieve optimal performance. According to the violin teacher Henry Saint-George, instructors needed to learn the "physical peculiarities" as much as the "mental characteristics" of their pupils.[88] In 1831, the doctor Francesco Bennati gave a lecture at the Royal Academy of Sciences in Paris, in which he attributed the Italian violinist Niccòlo Paganini's "prodigious talent" to "the peculiar conformation which enabled him to bring his elbows close together, and place them one over the other, to the elevation of his left shoulder, which was an inch higher than the right; to the slackening of the ligaments of the wrist, and the mobility of his phalanges, which he could move in a lateral direction at pleasure."[89] Books such as Anna Leffler Arnim's *Wrist and Finger Gymnastics* emphasized that instrument playing is "very arduous work," offering a series of daily arm, wrist, shoulder, and hand exercises for violinists to practice their bowing (see fig. 2.5).[90]

In this context, Rose's posture—particularly her raised hand and arm—mirrors the "peculiar conformation" of the body that Victorian thinkers like Lindsay and Bennati linked to sophisticated training and "prodigious talent." Rose is a skilled performer who has trained her body to play virtuosically. She does not display grotesque physicality or awkward bodily distortions

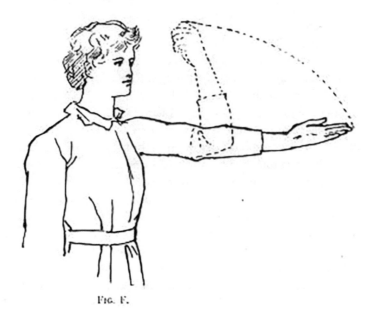

Figure 2.5. Exercise, in Anna Leffler Arnim, *A Complete Course of Wrist and Finger Gymnastics: For Students of the Piano, Organ, Violin, and Other Instruments*, 3rd ed. (London: Hutchings and Crowsley, 1894), 68. *Source:* Archive.org. Public domain.

but rather exudes corporeal power. She is quite literally well positioned to "astonish[]" her listeners.[91]

Violin playing also gives Rose access to the experience of joyous, kinesthetic motion. In one scene, she enters her room, "f[alls] first upon her violin, and rush[es] through a Brahms' 'Liebeslied,' her eyes dancing, her whole light form thrilling with the joy of it."[92] A few moments later, she "br[eaks] into a Strauss waltz, dancing to it the while, her cotton skirts flying, her pretty feet twinkling, till her eyes glowed, and her cheeks blazed with a double intoxication—the intoxication of movement, and the intoxication of sound."[93] Soon, she has to stop to catch her breath. Here, Rose's physical pleasure derives not from any kind of romantic entanglement or sexual desire but rather from the musical stimuli that she herself creates in private. For Rose, violin playing involves her entire body ("*whole* light form") and precipitates a series of dynamic movements ("fall," "rush," "thrill," "break").

Rose's musical arousal only intensifies as the novel progresses. Throughout the novel, she appears "quivering and relaxed," "flushed," and "excited," and her "nerves" are "all unsteady with music and feeling."[94] After one particularly vigorous performance, she lays down her violin, "her breast still heaving with excitement and exertion."[95] Again, Ward mobilizes the language of the nervous system to capture the simultaneous arousal and ease that Rose experiences when performing. While Victorian authors often use the language of flushed cheeks and heaving bosoms to describe romantic attraction, for Rose, it is *music making* that grants her the most erotic pleasure. Ward describes Rose's bodily gratification as neither licentious nor perilous but rather the "natural voice of her inmost self," explained by the science of sound.[96]

Toward the end of the novel, the plot of *Robert Elsmere* shifts—disappointingly, for many readers—from one centered on Rose's musical pursuits to one preoccupied with her failed courtship with Edward and her eventual marriage to the aristocrat Hugh Flaxman. *Robert Elsmere* does not go quite as far as some feminist readers might hope to grant its heroine a wildly successful solo career detached entirely from domestic life.[97] By the end, it seems that Ward, like the authors of the other "lady violinist" novels mentioned above, has accepted the conclusion of Rose's musical career in the face of marriage.

Yet, in the book's sequel, *The Case of Richard Meynell* (1911), Ward takes a crucial opportunity to revive Rose's musical life. Though readers see very little of Rose in *Richard Meynell*, when she does appear, it is in a musical context. Early in the novel, Rose complains that her husband is still chairman of the city council but insists, "I shall make him resign that, next year. Then we are going for six months to Berlin—that's for music—*my* show!"[98] Here, Rose exudes the kind of influence over her husband that she enacted over her listeners—and uses it to insist on her Continental musical pursuits. Though Rose does not, we are led to believe, become a Norman-Neruda figure, touring the world to solo with major orchestras to great global acclaim, Ward reassures readers that Rose's musicianship (rather than her marriage, motherhood, or other domestic obligations) remains her most treasured pursuit ("*my* show!"). While readers do not know exactly what those six months in Berlin will entail—Does Rose perform? Does she take violin lessons? Does she simply attend concerts?—six months abroad, especially in the nineteenth century, particularly for a female *violinist*, would in any case have represented a serious musical commitment. The "Brahms and Wagner in her young blood" cannot, perhaps, be so easily diluted.

"Itching" to Be Heard: *The Duenna of a Genius*

Whereas Ward merely hints at the persistence of Rose's musical life, M. E. Francis grants her heroine an unequivocally sustained musical career. Known mostly for her children's books, Francis was a lifelong devotee of music and sought in *The Duenna of a Genius* to make "music itself . . . [her] theme, and [her] characters . . . moulded by it."[99] Francis dedicated her "romance of music" to the Polish violinist Jan Ignace Paderewski and named each of the novel's chapters after musical moods or tempi, such as "Molto Espressivo" or "Staccato."[100] As she writes in the book's dedication, "I . . . would fain that the sound of music were heard the while my story is told."[101]

Like *Robert Elsmere*, *Duenna* centers on two sisters: the violinist Valérie and her sister Margot, who also serves as Valérie's accompanist, general manager, and handmaiden (the "duenna" of the novel's title). The two sisters move to London in hopes of advancing Valérie's violin career. At first, it seems *Duenna* will be a tale of a spoiled and self-important diva who ultimately fails to achieve acclaim among philistine and sexist British audiences. "Imperious," "irrational," and "spoilt," Valérie bosses Margot around, refuses to let anyone else touch her violin (which she herself handles with a "queenly air"), and (like Rose) scoffs openly at listeners who fail to appreciate her music.[102] As a result, the sisters struggle to gain a receptive audience for Valérie's playing. The aristocrat Lady Mary Bracken, for instance, hires the sisters to play at a party but asks Valérie to warm up upstairs because, she bemoans, she "can't endure the sound of a fiddle being tuned"—a request that Valérie (again, like Rose) brazenly defies as she "proceed[s] to draw forth a series of excruciating sounds."[103] Valérie's performance at Lady Mary's receives mixed reviews: "[One lady] murmured audibly that she liked something with more tune in it; while an old gentleman observed that he thought there was nothing like a banjo for a drawing-room—*his* girls were going to learn the banjo, he said. Valérie fixed her eye for a moment on Rosamond Gorst, who languidly returned the glance and proceeded to smother a yawn."[104] Francis's caricature of the audience's tactless and apathetic response—and their yearning for more appropriate drawing-room performances—suggests that *Duenna* might simply represent another "lady violinist" narrative about a woman unable to overcome either her own hubris or the musical prejudices of her world.

Yet *Duenna* ultimately resists the "diva" trope by emphasizing the physiological nature of Valérie's musicality. Valérie's musical skills are presented as instinctive, powerful, and undeniable—explainable by contemporary acoustical science and thus able surmount the social strictures that would otherwise limit them. Like Ward, Francis emphasizes that Valérie's music is deeply embedded in her body from a young age. Not only is Valérie referred to throughout the novel (including in the title) as a musical "genius," but

Margot also explicitly describes Valérie's musical skill as inherited: "[Our] father could play most beautifully; it is from him that Valérie inherits her genius."[105] In such moments, Francis rehearses the notions of Billroth, Gall, and others who believed that musical ability was a hereditary, inborn trait.

Valérie also experiences an overwhelming—"imperative" and "overpowering"—physical desire to play and be heard.[106] The narrator writes, "She had a message to deliver to the world, and she would know no rest until she obtained a hearing . . . no minor success would satisfy that craving."[107] After a performance for a particularly lukewarm audience, Valérie laments, "I think it will kill me—the longing to be heard and the maddening knowledge that no one wants to hear me."[108] Music is essential to Valérie's life force; it is a physical "craving" that she cannot help but doggedly pursue.

Valérie eventually wins over her philistine audience members with her undeniable musical aptitude and the physical power of her playing. When she performs for a concert at the Brackenhurst Town Hall, the "genuine music-lovers" in attendance applaud "rapturously" and "vehemently" and "make frantic endeavours to obtain a recall"—phrases that connote the intensity and urgency of their responses.[109] Elsewhere, her rendition of Beethoven's *Kreutzer Sonata* "float[s] through the room"—nodding to the physical ability of Valérie's music to travel through space—and appeals to all of her listeners regardless of musical aptitude: "There was no need to command silence now; curiosity had been aroused, and though the performance was of a very different order, its wonderful beauty appealed even to the most prosaic and unmusical person present. . . . At the close there was real enthusiasm and applause; people pressed round her with congratulations and admiring speeches."[110] Valérie's music is so wonderfully stirring that it sends her listeners into rapt silence despite themselves. While at first Valérie appears as a bit of a snob who scorns her uncultured audiences, here she emerges as an artist whose music exerts inevitable and universal effects on her listeners, so much so that they are physically drawn to her.

Valérie finds a particularly receptive listener in Sir John Croft, Lady Brackenhurst's nephew and Margot's future husband. When she plays a set of Russian variations at Lady Brackenhurst's dinner party, John finds himself utterly awestruck: "Croft had been prepared for something very good, but he had not expected anything quite so excellent as this. There was no doubt about it: Valérie Kostolitz was a great artist. He marvelled at the power of the little creature. What passion! What fire!"[111] John is awed not only by Valérie's artistic excellence but also by the physical energy and strength that she, even as a "little creature," displays as she performs. Her bodily involvement in her music does not strike him as problematic; rather, it leaves him with "no doubt" that she is a great artist.

John is even more arrested by Valérie's performance of the *Kreutzer Sonata* a few scenes later: "He was absolutely dumb: tears stood in his eyes; it seemed to him that those slender fingers of Valérie's were drawing the heart from out his breast. It was a very dream—never could he have conceived such perfection of tone with a charm so penetrating, so exquisite."[112] Several minutes later, Valérie notices "the emotion still evident in his face."[113] As much as this passage captures the metaphysical aspects of Valérie's playing—it seems to John as a "very dream"—it also hinges on descriptions of the concrete physical and physiological effects of Valérie's "penetrating" music, which brings tears to his eyes and even alters his physiognomy ("the emotion still evident in his face"). The image of Valérie's "slender fingers" on the strings draws attention to the tactility of her music making; in this context, her fingers "drawing the heart from out [John's] breast" reads less as a vague metaphor for musical influence and more as a concrete evocation of her haptic and sonic power over her listener's body.

As much as Valérie affects her listeners' bodies with her music, her playing ignites the most intense sensations in her own body as she performs. After one performance, both Valérie and Margot find themselves "trembling with excitement," reflecting the acute involvement of their nervous and muscular systems.[114] Even John observes the physiological pleasure Valérie enjoys while performing: "Her face was transfigured, her eyes dilated; she had even a majesty of bearing with which one could have credited her."[115] Just as Rose does, Valérie displays an active physical posture as she plays—one that suggests "magisterial" corporeal command rather than the inappropriate bodily contortions imagined by many nineteenth-century critics. Moreover, her eyes dilate, a phenomenon that nineteenth-century scientists were beginning to attribute to the arousal of the nervous system.[116] The German scientist Oswald Bumke wrote in 1911, "Every active intellectual process, every psychical effort, every exertion of attention, every active mental image, regardless of content, particularly every affect just as truly produces pupil enlargement as does every sensory stimulus."[117] Here, it is Valérie who creates the "sensory stimulus"—the music—that invigorates her nervous system.

Valérie's physiological ties to her music also render her a passionate and active listener. When John takes Valérie to hear a famous violinist's recital at St. James's Hall, she cannot help but physically participate in the performance, even though she is not onstage herself: "It was amusing and exhilarating to watch the little creature's pleasure. . . . He felt it to be a privilege to watch her absorbed face, marking how the play of the mobile features varied with the changes in the music. . . . He could even see her hands working, the fingers curving themselves involuntarily, as though they too itched to handle bow and strings."[118]

Valérie's actions here echo contemporary scientific discourses about musical muscle memory, which understood musical actions such as fingerings and bow strokes as indelibly rooted in the nervous systems of those who played extensively and thus able to be automatically recalled, even subconsciously. George Henry Lewes, for instance, argued that learning to play a musical instrument involved the formation of "Habits, Fixed Ideas . . . Automatic Actions" or "muscular contractions" that could be repeated and "performed while the mind is otherwise engaged."[119] Similarly, the physician H. Hayes Newington said that "a great portion" of the "mental operations" involved in musical performance "must be done by reflex action" and "without conscious interference on the part of the higher centres."[120] Newington describes hallucinating and otherwise mentally ill patients who can play "difficult music" due to the "reflex excitation of the motor centres," even while a "storm" is going on in their brains.[121] Herbert Spencer also theorized that musical practice results in physiological changes that enable the "practised musician" to play "while his memory is occupied with quite other ideas than the meanings of the signs before him."[122]

In this context, readers can see how Valérie's fingers are indeed primed for performance, eager to curve around an imaginary violin. This process is "involuntary," rooted in the unconscious and automatic actions of her nervous system. Despite that Valérie's violin is not physically in her hands, her musical muscle memory enables her to take part in the performance. Her musicality is so deeply embedded in her body that she is able to automatically reproduce it even when there is no instrument present. Valérie is not the passive female listener seen elsewhere in Victorian fiction; she is an active and engaged participant in musical creation.

At times, Valérie's obsession with music becomes so intense that others worry that it might result in her nervous collapse. Echoing the ideas of thinkers like Lunn and Tyacke George, who believed that women lacked the nervous strength required for music making, Margot wonders whether Valérie's "delicate artistic brain and highly strung nervous organization [are] giving way under the pressure of this sudden fierce excitement"—implicitly emphasizing the embeddedness of music in Valérie's body by likening her nerves to strings.[123] Valérie's doctor diagnoses her with "nervous exhaustion" because she has "overtaxed" herself, and Margot takes her to Wiesbaden,

Germany, to recover.[124] Yet, despite these attempts to medicalize her and limit her music-making—Margot even locks up her violin—Valérie rejects the notion that it was her music that exhausted her, refusing to "eat, drink, sleep or go out until her treasure [her violin] was restored to her."[125] In fact, the deprivation of music does *more* harm to Valérie's nervous system than does its excess: "The external quiet of nature roused a kind of frenzied impatience in Valérie. She was not quiet; the blood was leaping in her veins; she felt a passionate need of movement, of action; she wanted above all to play."[126] Just as Rose feels "Brahms and Wagner in her blood," Valérie experiences her desire to play as a physiological need embedded in her blood and veins.

Valérie's physical need is so strong that it again compels her to "play" even without an instrument: "Valérie sat down on the windowsill, drumming, with unquiet, impatient fingers, on the panes. Was it not so the *Rêverie* went? She could remember it note by note."[127] As when she "itches" to play in the concert hall, she activates her muscle memory to gain access to the music she so desperately needs, transforming the windowpanes into a surrogate instrument. When Margot is forced to give in and return the violin to her sister, Valérie "play[s] constantly." Her playing, combined with the "fresh air, sunshine, change of scene, and beautiful creamy milk" in Wiesbaden, physically restores her: "Her small hands were growing plump again, the colour was coming back to her face."[128] Music, like temperate weather and nourishing food, is curative rather than ruinous to Valérie's body.

As the novel progresses, Francis introduces an extensive romance plot between Valérie and a Hungarian pianist named Paul Waldenek. Yet the author explicitly distances this storyline from most of the courtship narratives found in other "lady violinist" novels by basing Valérie and Paul's romance on their deeply physical—and, crucially, mutual—musical exchanges and by insisting that their relationship facilitate rather than inhibit Valérie's musical career. Throughout the narrative, Francis makes clear that this romance is mostly about the music. In fact, Valérie's first encounter with Paul is not with the man at all but with his music. When Valérie attends one of his concerts, she "bolt[s] upright, her eyes dilated, her lips parted," and "trembl[es] so violently."[129] This episode serves not merely as a vague metaphor for Valérie's budding romantic attraction to Paul but also as an overt description of the palpable effects of his music on her body. Yet she is far from a passive listener, moved helplessly by her suitor's music. Valérie again participates kinesthetically in the musical performance; she "beat[s] her little hands and str[ikes] her feet frantically on the floor," throwing her whole body into the performance even when she is not the one onstage.[130]

Valérie returns home from Paul's concert not passively musing about the handsome violinist but rather brimming with creative inspiration and in hot pursuit of her own musical work. She immediately retrieves her violin to play Paul's piece herself. Importantly, though, she does not merely reproduce his music but seeks to improve upon it. With "burning cheeks," Valérie exclaims to Margot, "His fingers were wonderful; but my Cremona can say more than any piano. Listen!"[131] Valérie is able to replicate what she has heard (further evidence of her sophisticated musical memory)—and, crucially, improve it. Even Margot is "thrilled by unwilling admiration" as she listens to Valérie's adaptation: "When she came to his own compositions she threw into her rendering of them such tenderness and yearning passion that tears rushed to Margot's eyes."[132] Here, readers further sense that it is Paul's music, rather than him per se, that most excites Valérie. The music invigorates her not because it is so brilliant in and of itself but because she can make it better.

Valérie's passion for Paul's music is not one-sided; Paul is just as physically entranced by Valérie's playing as she is by his compositions. One evening, she hears footsteps outside of the house and, upon realizing they are Paul's, seizes her violin and rushes across the street to follow behind him unseen. When Paul stops to rest on a wooded path, Valérie begins to play his *Rêverie*:

> Of a sudden, music fell upon his ear—his own music, his *Rêverie*, played with wonderful tenderness and expression on the violin. His first and predominant sensation was that of surprise—surprise, not at the unusualness of hearing music in such a place and at such an hour, but that his *Rêverie*, written for the piano, should adapt itself so exquisitely to the violin. He listened spellbound, the beauty of the theme—his theme, conceived by his own brain, his own heart—intoxicating him as it had not done even in the first ecstasy of composition. He was carried away by his own passion, uplifted by his own desire. Tears stood in his eyes, and yet he smiled. Then all at once it was borne in upon him that the unseen musician was an artist, more than an artist, a genius. Only a genius could give evidence of such sympathy, such intuition, such extraordinary power.[133]

Francis's depiction of this musical exchange certainly abounds with metaphysical language, punctuated by words like "uplifted," "carried away," and "spellbound," as well as the title of the piece ("*Rêverie*") and Paul's perception that "his own soul and hers were beseeching Heaven."[134] Gillett suggests that this scene reflects Valérie's "almost magical" wooing of Paul.[135] Yet I argue that this moment is as physiological as it is mystical. The image of the music falling upon Paul's ear reminds us that this is a sense-based encounter, based on Paul's reception of Valérie's music through a bodily organ. Paul's sensual arousal is further punctuated by him smiling and crying, references to his brain and heart, and words like "intoxicating" and "ecstasy." Valérie's "exquisit[e] adapt[ation]" of Paul's music—a detail that further highlights her musical "genius"—fills him with "passion" and "desire." Importantly, he feels even more "intoxicated" by *her* playing than by his own composition. This scene thus reverses the common trope of the passive female listener enthralled by a male genius. Here, Valérie is the player, igniting Paul's senses with her improvements to his music.

The narrator makes clear, however, that Paul's initial passion is more musical than romantic, as he does not learn of Valérie's identity until after she finishes playing. He is so struck by the "genius" of the player he hears, in fact, that he erroneously assumes the player is a "he"—both nodding to the rarity of Valérie's skill and implicitly critiquing common Victorian assumptions about women's musicianship as inevitably inferior.[136] By maintaining Valérie as an "unseen musician," Francis assures readers that Paul's physical reaction is based on the music itself rather than its ties to the performer as romantic object.[137]

As the romance between Valérie and Paul develops, music provides the vehicle for their budding romance and serves as the central nexus for their interactions and shared physiological sensations. Valérie's story reveals an instance in which, to quote Przybylo, "the erotic fuels sexual desire rather than sexual desire being at the base of the erotic."[138] Indeed, for Valérie and Paul, the erotic sensations of music making and listening take priority over any kind of sexual relationship. When Margot asks Valérie about one of her conversations with Paul, Valérie replies, "I do not think we talked at all. I played—we talked of music."[139] Similarly, the narrator later writes, "Their talk was chiefly of music; scarcely had they met before the instruments were in requisition; they wooed each other with exquisite sounds."[140] The couple's musical collaborations ignite physiological sensations in both themselves and their listeners: "Valérie had played as she had never played before; Waldenek had played with her—they had been lost to all sense of their surroundings; even Margot, leaning back in her corner, had become almost as excited as they. . . . She was still breathlessly marvelling and admiring."[141] For Francis, the only acceptable condition under which Valérie could have a romance is if it enhanced her playing and her ability to physically affect others with her music.

Though Valérie and Paul eventually marry, their marriage launches rather than inhibits Valérie's musical career. Francis insists that Valérie's music remain the couple's primary focus. Unlike John, who tells Margot "I should not like you to play in public once you belonged to me," Paul makes it his mission to promote Valérie's musical career.[142] Once Paul helps Valérie get onstage, her success is "immediate and complete. It could scarcely be otherwise. . . . The enthusiasm of the audience knew no bounds. Indeed, no such musical treat had for years been offered to the public."[143] Although some readers might be disappointed that it takes a man to enable the ultimate flourishing of Valérie's violin career—Margot herself laments, "*This* was not the success they both had planned; this was not Valérie Kostolitz taking the world by storm entirely by the power of her own genius"—Francis assures readers that Valérie's career ultimately takes priority over Paul's.[144] Rather than fostering his own solo career at the expense of hers, Paul plays the piano accompaniment for Valérie's public performances—a reversal of the more traditional Victorian musical setup in which a female pianist accompanies a male performer. After a performance at St. James's Hall, "even Margot was satisfied with the manner with which Waldenek subordinated himself to Valérie, his performance blending itself with hers in a manner which sustained without ever dominating it."[145] At the end of *Duenna*, then, readers are reassured that Valérie's deepest physiological desires will be fulfilled—not because she finds romantic attachment but because she will be able to continue her music. Francis suggests that female violinists can have romances, but they must be based on the pursuit of music above all else, for it is music that will best nourish their bodies and enable them to impact the world around them.

Physiological Acoustics and Feminist Politics

To twenty-first-century feminist readers versed in theories of antiessentialism and gender performativity, it is perhaps unsettling that the feminist interventions of *Robert Elsmere* and *Duenna* hinge so emphatically on the physical body. As Elizabeth Wilson writes, the body is often seen as "the underbelly of feminist theory . . . a dank, disreputable mode of explanation and a site of political vulnerability."[146] And yet, as several feminist and queer science theorists have argued, the body need not always be antithetical to radical gender politics.[147] Elizabeth Grosz asks, "What are the virtualities, the potentialities, within biological existence that enable cultural, social, and historical forces to work with and actively transform that existence? How does biology—the structure and organization of living systems—facilitate and make possible cultural existence and social change?"[148]

In their use of physiological acoustics to imagine new and vibrant possibilities for their heroines' phenomenal lives, *Robert Elsmere* and *Duenna* locate the "potentialities" of biological science for social change. These novels remind readers—especially those who are necessarily wary of essentialist thinking and suspicious of appeals to science to explain embodied experience—that a focus on the materiality of the body can at times be mobilized to defend those for whom the world of social relations offers little sustenance.

Chapter Three

Cross-Dressing Violinists and Music/Gender Performance in *The Heavenly Twins* and *The Violin-Player*

In Bertha Thomas's 1880 novel *The Violin-Player*, a female violinist named Laurence Therval performs in public. She plays with passion and vigor, intently focused on her music: "At such moments the outside world was almost non-existent for the little musician. Laurence had early acquired the power of thus throwing all the nerve-force in her—so to speak—into one channel as she played."[1]

In many ways, this scene recalls those discussed in chapter 2, in which Rose and Valérie relish the pleasures of performance and stimulate their listeners' bodies with their music—and in doing so challenge sexist Victorian rhetoric about female violin playing. Laurence, too, displays a dynamism and virtuosity that flies in the face of misogynist diatribes about female musicality. As she activates her arm muscles and recruits her "nerve-force" (a strikingly Helmholtzian term), she demonstrates the distinctly physiological power of her playing.

Yet Laurence's story diverges from Rose's and Valérie's in a crucial way: Laurence does not always perform *as* a woman. A violin prodigy from early childhood, Laurence moves to Berlin as a young adult to audition for the virtuoso instructor Professor Nielsen. After she inquires at his studio, however, she receives a letter stating that Professor Nielsen does not—and will never—accept "lady pupils."[2] Echoing Victorian society's unofficial "ban" on female violinists, Nielsen's letter urges Laurence to give up the violin and focus on more acceptable feminine musical pursuits such as piano and voice.³ Instead of heeding her friend Linda's gentle suggestion that she might learn the pianoforte—the violin is, after all, "rather an awkward, unusual instrument for a lady to play"—Laurence chops off her hair and exclaims, "Shouldn't I do for a boy, now?"⁴ She determines that she does not even need to change her name because "Laurence" can refer to either gender; it is a boy's name in England and a girl's name in France. Laurence knows that if she can pass as a boy, a world of musical opportunity will open to her. And indeed, Nielsen swiftly admits the male Laurence into his studio, struck by the genius of the "little fellow."⁵

This chapter explores the recurring figure of the cross-dressing female violinist in two late-nineteenth-century novels: *The Violin-Player* and Sarah Grand's *The Heavenly Twins* (1893), in which the violinist Angelica Hamilton-Wells dresses as a man to enter the studio of a local church musician and play with him. Whereas Rose and Valérie openly resist the masculinist codes of the Victorian musical world by pursuing professional musical careers, Angelica and Laurence attempt to sidestep such codes altogether by performing as men. While *Robert Elsmere* and *The Duenna of a Genius* assert the virtuosity and power of female musicians, *The Heavenly Twins* and *The Violin-Player* explore what happens when the category "female" falls away. Cross-dressing allows Angelica and Laurence to fully enter the realm of professional musical performance and, at least temporarily, escape the kinds of artificial gender rules that would otherwise relegate them to the drawing room or subject them to social scorn. Dressed as men, Angelica and Laurence are able to move freely through the musical world and showcase their skills, which are equal to—and often surpass—those of their male counterparts. Both novels thus put a fine point on the hypocrisy of Victorian music critics such as Sutcliffe or Honeyman, who denigrated female musicianship as inherently subordinate. It is their musical prowess, in fact, that helps Angelica and Laurence to "pass" in the first place; neither Professor Nielsen nor the church musician would ever expect that a woman could display such rapturous virtuosity.

At first blush, these cross-dressing narratives resonate obviously with twentieth- and twenty-first-century theories of gender performativity, most notably Judith Butler's now-canonical claim that gender is achieved through a "stylized repetition of acts."⁶ After all, both *The Heavenly Twins* and *The Violin-Player* show just how easy it is for a woman to pass as a man by cutting her hair and donning different clothes. Angelica outlines her relatively straightforward process of gender transformation:

> I saw a tailor's advertisement, with instructions how to measure yourself; and I measured myself and sent to London for the clothes—these thin ones are padded to make me look square like a boy. And then, with some difficulty, I got a wig of the right colour. It fitted exactly. . . . But isn't it surprising the difference dress makes? I should hardly have thought it possible to convert a substantial young woman into such a slender, delicate-looking boy as I make.[7]

According to Angelica, slipping in and out of genders can be achieved through simple processes of measuring, purchasing, and dressing. Similarly, when Laurence chops off her hair and exclaims, "Shouldn't I do for a boy, now?," she intimates that "doing for" is all that is required.[8] In carrying out such acts—cutting their hair, donning wigs, dressing in "thin" and "square" clothes, and manipulating their voices—Angelica and Laurence demonstrate that masculinity can be performed relatively easily. "Dress," as Angelica remarks, can make all the "difference."

And yet, the presence of acoustical science in these cross-dressing scenes complicates such a neatly performative reading. Always hovering over these moments of gender manipulation are insistent reminders of the characters' musical bodies—their quivering limbs and twitching muscles, their "nerve-force" and muscular power. While Angelica and Laurence are nonchalant, even flippant, about their acts of gender manipulation, they are ruthlessly committed to their music, which they experience as an essential, visceral need. The novels themselves also prioritize music over gender at a narrative level. Both works focus in much greater detail on the corporeal sensations aroused by music making than on the embodied experiences of living as a particular gender. In *The Heavenly Twins*, Angelica also cross-dresses for the *readers*, who do not know she is the Boy until the very end of the story. Readers are thus prompted to focus on the viscerality of the Boy's music making rather than on the cross-dressing narrative. In *The Violin-Player*, while readers do know that the "boy" Laurence is in fact a "girl," the ultimate reveal of her gender is, shockingly, no big deal in the context of the novel. Laurence has by then so thoroughly convinced Nielsen of her innate musical ability—and overwhelmed *his* body with her playing—that everyone involved moves on rather swiftly from the topic of her gender transgression. Both novels suggest that while gender may be manipulable, musicality is innate.

In their framing of musicality as essential, then, *The Heavenly Twins* and *The Violin-Player* actually lie in tension with much feminist and queer theoretical work on gender performativity. For Angelica and Laurence, musicality provides the kind of innate bodily "core" that theorists like Butler, at least in *Gender Trouble*, famously resist.[9] The essentialist language in *The Heavenly Twins* is especially unnerving given Grand's own ties to eugenics and biological determinism, including her insistence on sex difference and rational reproduction.[10] And yet, both novelists draw on this paradoxical fusion of performative and essentialist frameworks in order to articulate their own kinds of gender politics. Where Ward and Francis insist on reclaiming the unique powers of patently female performers, Grand and Thomas explore the possibilities that emerge when women are able to, however temporarily, set aside their gender altogether. Angelica and Laurence seek to become not *women musicians* but simply *musicians* (a privilege granted automatically, of course, to male musicians). By momentarily shedding their female genders, Angelica and Laurence access their most gratifying, erotic sensations of nourishment, invigoration, relaxation, pleasure, and arousal—sensations they feel at the deepest levels of their muscles and nerves. Grand's and Thomas's gender politics thus hinge not on celebrating the power of the female musician but on insisting that some things—including music—are more important, more essential to the body, than gender. The most daring intervention of these works, perhaps, is that they depict women who are fundamentally more committed to music than to any sense of gender identity.

This notion of a rich bodily life apart from—or alongside—particular forms of gendered existence would have been a powerful provocation at a time when women were defined, above all, by *being women*, and when women's bodily fulfillment was supposed to be attached solely to their rehearsal of gendered roles (marriage, marital sex, reproduction, motherhood). Even today, narratives about women's "biological clocks" or mothers' "natural instincts" (indeed, the very ideas, still maintained by many politicians and trans-exclusionary feminists, that mothers *are* always women and that reproductive rights are inherently "women's rights" rather than the rights of "pregnant people") still persist in many Western societies, suggesting that a woman's bodily life is governed, above all, by her gendered being.[11] Yet, in their cross-dressing violinist narratives, Grand and Thomas are able to ask: What else besides "womanhood" might contribute to a rich, erotic existence?

In posing this question, *The Heavenly Twins* and *The Violin-Player* resonate less with Butlerian theories of gender performativity and more with recent work in queer, trans, and asexuality studies that makes space for forms of eroticism and intimacy that, Przybylo writes, "are simply not reducible to sex and sexuality."[12] In particular, theorists of queer and trans musicology such as Taylor and Baitz suggest that scholars can attend to the viscerality of musical response while still retaining an emphasis on the fluidity of gender and sex. As Taylor and Baitz both argue, the commitment to antiessentialism among queer musicologists has at times resulted in the erasure of the material dimensions of music listening and performance and of music's potential as a "queer erotic reality beyond the boundaries of gender, sexed bodies, and specific bodily orientations."[13] Similarly, Baitz notes that music can be carnal without being essentially gendered or sexual: "Musical reception can be described without downplaying the influence of notes and sounds. . . . The fact that the music is experienced quite carnally has no relation to the way that gender identity is signified. . . . The visceral response (or lack thereof) that music can elicit has no bearing on how sexual identity is produced, the kinds of identities facilitated by the music, or the methodologies best suited to investigating it."[14]

The Heavenly Twins and *The Violin-Player* navigate similar debates (albeit in different terms) about the tensions between musical perform*ance* and gender perform*ativity*. These novels invite readers to consider: When are gender and sex useful categories of identity, affiliation, and coalition, and when should the body be theorized *beyond*, or *alongside*, these terms? What if the body is not only (or not always) gendered or sexual but *musical*? Does music—especially in the context of physiological acoustics—allow for discussions of corporeal pleasure and gratification outside of strictly gendered or sexual terms? What opportunities might arise when writers and thinkers make room for other "essential" bodily experiences?

Grand's and Thomas's novels differ in the extent to which they ultimately subsume their heroines back into the Victorian gender systems they temporarily departed. The two writers, it turns out, have very different answers to what happens when their heroines' female genders are reinstated. Angelica's access to musical pleasure is entirely contingent on her cross-dressing. She can only fully embrace her musical body when she is dressed as "the Boy." Her musical pleasure—even her very ability to play—fizzles when she reveals her female identity and is relegated back to the drawing room. Grand thus casts music's physiological power as situational and transitory; Angelica's body, innately musical though it may be, cannot fully escape the social codes that dominate Victorian women's lives. *The Violin-Player*, on the other hand, presents a far more radical outcome. Unlike Angelica, Laurence shatters the prohibitions that once limited her and enjoys a successful musical career even after she reveals her female identity. Music teachers, critics, and listeners are so in awe of Laurence's musicality—and, crucially, understand its undeniable physiological effects—that they permit and even celebrate her pursuits. However divergent their conclusions, though, both novels offer new frameworks for forms of erotic gratification that can occur without gender as a stable—or stabilizing—force. For Angelica and Laurence, it is not by marrying or giving birth, nor even by celebrating their uniquely "female" musical powers, but by chopping off their hair and seizing their violin bows, that they are able to access their most profound erotic sensations.

"The Difference That Dress Makes": Contingent Virtuosity in *The Heavenly Twins*

The cross-dressing narrative in *The Heavenly Twins* is notable not only for its depiction of gender subversion but also for its inclusion in the novel in the first place. The episode constitutes its own "book"—Book IV: "The Tenor and the Boy: An Interlude"—situated somewhat awkwardly in the middle of the novel. Grand originally wrote "The Tenor and the Boy" as a separate short story, which she unsuccessfully attempted to publish on its own in 1890.[15] Book IV's events have little to do with the rest of the novel; the narrator refers to the main characters only by their aliases (Tenor and Boy), neither of whom readers have encountered—or think they have encountered—before, as they do not learn that the Boy is Angelica until the end of the episode. Because of its seeming disjunction from the rest of the novel, critics have questioned whether the episode should have been included at all.[16] Grand herself even admitted that she thought "The Tenor and the Boy" might not have been entirely "necessary to the story."[17]

Out of place though it may be, "The Tenor and the Boy" offers some of Grand's most striking explorations of gender, the body, and performance. While Angelica is not Grand's first musical heroine—female musicians populate the pages of *The Beth Book* (1897), *Ideala* (1888), and "Mama's Musical Lessons" (1878)—she is the only one to cross-dress in order to perform and thus the one who achieves the most expansive, erotic musical life. Angelica's successful gender deception allows her to delight, however fleetingly, in an artistic world that is otherwise unavailable to her.

A plucky, precocious heroine in the vein of *The Mill on the Floss*'s Maggie Tulliver or *Little Women*'s Jo March, Angelica chafes against gendered limitations from the very beginning of the novel. Due to her independence, her tomboyishness, and her feisty temperament, critics often associate Angelica with the New Woman movement of the late nineteenth century, in which Grand was involved.[18] Unsurprisingly, most of Angelica's attempts to resist Victorian gender norms—including her efforts to join her twin brother Diavolo's activities and her proposal to her husband, Mr. Kilroy of Ilverthorpe—are met with shock and scorn. In a scene that prefigures Angelica's cross-dressing in Book IV, she and Diavolo exchange clothes for a wedding, and the mother of the bride, horrified, insists that their actions are "unnatural" and would "bring bad luck."[19]

Only by dressing as the Boy can Angelica achieve her most significant gender transgression. Early in Book IV, a professional church musician called "the Tenor" comes to Angelica's town of Morningquest and arouses the fascination of the whole community with his vocal talents. Entranced by the Tenor's music—and, readers are later led to believe, eager to flirt with the Tenor and find out what he thinks of her—Angelica dresses as the Boy and begins to visit the Tenor in his studio at night. Though her aims may have initially been more romantic than musical, soon it is the nightly opportunity to play her violin that most gratifies Angelica: "You see, I loved to make music. Art! . . . I wanted to *do* as well as to *be*."[20]

Cross-dressing not only grants Angelica logistical opportunities to play but also enables her to access the physiological gratification that music making provides. While Angelica's cross-dressing may work to "deny the patriarchal insistence that a woman is always already her body," as Martha Vicinus writes, I propose that it also enables Angelica to satisfy her visceral need to make music.[21] Grand showcases the bodily strength and pleasure that Angelica senses while performing and captures the autonomic responses that her music induces in her listeners. The syphilitic and pregnant bodies elsewhere in the novel highlight the dangers of physical (especially sexual) life, whereas the musical bodies in Book IV vibrate with aesthetic pleasures and as-yet-unrealized erotic possibilities.

Playing violin as the Boy allows Angelica to discover new features of her body, such as its capacities for vigorous output and free, unhampered movement: "He took up his violin and played a plaintive air, to which he chanted . . . executing . . . steps himself at the same time in illustration."[22] Later, we see the Boy "rejoice in his own strength, to delight in his own suppleness; and he walked on now with healthy elastic step . . . the bow in his hand, now flourishing it in the air, and now drawing it across the instrument."[23] As the Boy, Angelica is able to enact the unrestricted physical movements necessary to play the violin (movements that were, of course, off-limits to Victorian women). Angelica-as-the-Boy savors the way in which violin playing requires the involvement of the whole body (feet, arms, hands, and voice) and enables exaggerated kinesthetic actions (flourishing the bow, drawing it across the instrument, and stepping in time to the music). Her "suppleness" and "elasticity" not only nod to the plasticity of her gender but also highlight the dynamism, agility, and flexibility of her body—its limberness and litheness, its ability to expand and contract, and its capacity to take up space and embrace its impulses. The music compels Angelica's muscles to move, and, dressed as the Boy, she can embrace these compulsions.

Angelica-as-the-Boy produces her own vocal accompaniments for her violin performances. As she plays and sings, the Tenor ponders, "That voice of his was wonderfully flexible; he could make it harsh, grating, gruffly mannish, and caressing as a woman's, at will, but the tone that seemed natural to it was the deep, mellow contralto into which he always relapsed when not thinking of himself."[24] Angelica experiments with all of the vocal flexibility her body has to offer, manipulating her mouth, throat, and lungs to emit a range of sounds. As the Tenor's musings make clear, her vocal experimentation is also a form of gender play. While a higher, softer pitch would have been expected of female vocalists during the Victorian period, Angelica-as-the-Boy can try out both "mannish, and . . . woman[ly]" sounds (with the article "and" even suggesting their simultaneous occurrence). Angelica ultimately lands on a "deep, mellow" contralto—a range that is "not-bass, not-soprano," according to musicologist Wendy Bashant, and thus occupies a liminal space between male and female vocal ranges.[25] The contralto range is not simply novel or interesting to Angelica; rather, it is the voice that is most "natural to" her, the one to which she automatically "relapse[s]"—a word that signals a return to an original state. Dressed as a man, Angelica can most freely harness her essential bodily sensations.

Playing for the Tenor also provides Angelica with a fundamental source of bodily sustenance. In a rather on-the-nose nod to the nourishment her music provides, the narrator explicitly pairs the act of music making with that of eating: "When [the Boy] was tired of making music, as he called it, he demanded food, and, so long as he could cook it and serve it himself, he delighted in bacon and eggs, as much as he did in Bach and Beethoven."[26] The parallel of Bach and Beethoven with bacon and eggs foregrounds music as a form of sustenance—and here, dressed as a man, Angelica can finally partake. While her childhood cross-dressing was deemed "unnatural," here, cross-dressing enables Angelica to satisfy her most natural human needs.

Angelica-as-the-Boy revels in the physical influence of her music on both things and people. When she plays violin, she "scarcely seem[s] to touch the strings, yet wak[es] low Æolian harplike murmurs, or deep thrilling tones, or bright melodious cadences; making [the violin] respond to his touch like a living creature."[27] The abstract, metaphysical language that abounds in this passage—particularly with the reference to the aeolian harp—appears alongside more concrete references to the physical effects of Angelica's music.[28] Her touch is not absent but light ("*scarcely* seem[s]") and evidently quite powerful. In the context of Helmholtz's theories of sympathetic vibration, the "murmurs" and "thrilling tones" created by her fingers highlight the material effects of her music. The image of the violin as a "living creature" casts Angelica not only as a kind of snake charmer or animal trainer who commands obedience to her will but also as a musician who can enliven all kinds of animate worlds.

Angelica awakens responses in her listener's body as well as the strings of her violin. She relishes the chance to affect the Tenor's nervous system: "I'll make you respect these delicate fingers of mine. . . . I'll make you quiver."[29] These future-oriented declarations reveal Angelica's eagerness to produce music that will penetrate the Tenor's ears and activate his nerves and muscles. Angelica-as-the-Boy indeed achieves these corporeal effects: "It was to his senses absolutely, not at all to his intellect, that the Boy's playing always appealed; but he did not quarrel with it on that account, for music was the only form of sensuous indulgence he ever rioted in, and besides, once under the spell of the Boy's playing, he could not have resisted it even if he would, so completely was he carried away."[30] As much as this passage makes use of metaphysical and even magical terminology (the "spell" of the music), it also recapitulates contemporary physiological treatises about music's primarily sensuous influence. The Tenor's musical responses are located in his body, not his brain, and his "sensuous . . . riot[]" is unwavering ("always") and unavoidable ("could not have resisted"). Though his responses may not be overtly sexual (he remarks at several points that he finds the Boy distinctly sexless), they are undeniably sensual.[31]

At the end of Book IV, Angelica's cross-dressing illusion shatters, along with the Tenor's perception of her genius and the feelings of bodily sustenance and strength that music making once granted her. One night, when they go rowing, the Boy falls out of the boat. As the Tenor rescues the Boy, both the Tenor and readers are shocked to learn that the Boy is in fact a woman, Angelica. The Tenor can no longer treat the Boy's once-charming gender ambiguity with the same bemused distance once he realizes this. After Angelica reveals her identity, the Tenor's attitude toward her shifts immediately and dramatically. His confidence in her bodily strength evaporates; he gives her "half the quantity of brandy he would have used five minutes before for the boy," despite the fact that, as the Boy, she had displayed the ability to eat and drink with aplomb.[32] The Tenor ponders the situation: "It was only a change of idea really, the Boy was a girl, that was all; but what a difference it made. . . . At any other time the Tenor himself might have marveled at the place apart we assign in our estimation to one of two people like powers, passions, impulses, and purposes, simply because one of them is a woman."[33] The Tenor's reaction conveys just how necessary Angelica's cross-dressing was. Though he is aware that his changing attitude is based on a shift of "idea" rather than reality, he cannot help but register the "difference" it makes. The conditional phrase in this passage ("might have marveled") reveals that though the narrator recognizes that women might have the same "powers, passions, impulses, and purposes" as men, the Tenor cannot help but see Angelica in any context other than her "assigned" place.

Angelica, too, knows that the Tenor will never again see—or hear—her in the same way now that he knows she is a woman. She tells him, "I have enjoyed the benefit of free intercourse with your masculine mind undiluted by your masculine prejudices and proclivities with regard to my sex."[34] For Angelica, the loss of this "free intercourse" is particularly devastating because it had given her brief access to bodily freedoms that felt "natural" to her: "It came naturally; and the freedom from restraint, I mean the restraint of our tight uncomfortable clothing, was delicious. I tell you I was a genuine boy. I moved like a boy, I felt like a boy; I was my own brother in very truth. Mentally and morally, I was exactly what you thought me."[35] Dressing as a boy *made* Angelica a boy; her gender performance gave her the license to move freely, delight in her body, and explore its strengths as she played the violin.

The novel soon relegates Angelica back to her "assigned" (rather than her "natural") role in Book V, "Mrs. Kilroy of Ilverthorpe"—a title that itself subsumes Angelica back into the context of normative domesticity, her identity hinging on that of her husband and his estate. She protests her domestic confinement musically:

> She got a violin and began to tune it. She was too good a musician not to be able to make the instrument an instrument

> of torture if she chose, and now she did choose. She made it screak; she made it wail; she set her own teeth on edge with the horrid discords she drew from it. It crowed like a cock twenty-five times running, with an interval of half a minute between each crow. It brayed like two asses on a common, one answering the other from a considerable distance. And then it became ten cats, quarrelling *crescendo*, with a pause after every violent outburst, broken at well-judged intervals by an occasional howl.[36]

Angelica retains some agency here, harnessing her musical skills to produce sonic torture. "Too good a musician not to," she creates whole worlds with her violin playing, transforming the drawing room into an animalistic frenzy. Her "creatures" run wild, paralleling her own agitation.

Soon, though, deprived of opportunities for uninhibited performance, Angelica struggles to channel her musical abilities. She appears

> stumbling over an air . . . a dismal minor thing which would have been quite bad enough had she played it properly, but as it was, being apparently too difficult for her, she made it distracting, working her way up painfully to one particular part where she always broke down, then going back and beginning all over again twenty times at least, till Mr. Kilroy got the thing on the brain and found himself forced to wait for the catastrophe each time she approached the place where she stumbled.[37]

Though Angelica is a musical "genius" when she performs as the Boy, as Mrs. Kilroy, she is unable to access her most basic musical skills. Despite repeated practice ("twenty times at least"), she is unable to master the difficult passage. The bodily pleasure and sustenance she once enjoyed have been replaced by labored, torturous, "painful" work. In a later scene, she sings the accompaniment to a series of piano exercises but is again unable to "produce it with the desired effect" because, the narrator reveals, she has not eaten enough that day.[38] Her physical starvation parallels her spiritual poverty—a stark contrast to the scene in which she devoured the bacon and eggs as the Boy. Absent of her male clothes—and the opportunities for unencumbered music making that they granted her—Angelica struggles to access her physical vigor: "In spite of her superabundant vitality, she had lost all zest for anything."[39]

At the end of the novel, Angelica somberly ponders how "her hour" is over. She learns that the Tenor has died as a result of an illness he contracted from jumping in the water to save her—a rather heavy-handed reminder of the dangers of "gender trouble." Angelica begs her husband to forgive her, attaching her apology to a promise to "never play in public as long as I live."[40] Grand even denies Angelica the opportunity for more modest musical pursuits; her husband does not permit her to "support a charity hospital with [her] violin," even though philanthropy was a common musical outlet for upper-class Victorian women.[41] As she puts away her violin and kneels at her husband's feet to beg for forgiveness, Angelica transforms, Ann Heilmann writes, "from free spirit to tame child-wife."[42]

The Heavenly Twins thus leaves readers with the unsettling notion that, though Angelica possesses innate musical virtuosity, the deeply entrenched social perceptions of women as inferior artists are even more powerful. Angelica's physiological ties to music are strong, but not strong enough to overcome the force of social norms. The music is somewhere in her fingers and muscles, but she can no longer access it. As her husband tells her, "The influences of sex, once the difference is recognized, are involuntary."[43] Though Angelica is able to glimpse new bodily experiences when she plays as the Boy, she ultimately recognizes that it is impossible for a woman to realize and sustain such experiences and for others to resist their "involuntary" assumptions based on gender.

In this context, the epigraph to Book I, a statement from Darwin's *Autobiography* (1887), reads as tragically ironic: "I am inclined to agree with Francis Galton in believing that education and environment produce only a small effect on the mind of anyone, and that most of our qualities are innate."[44] Darwin's statement, troublingly informed by Galton's scientific racism (and likely quoted due to Grand's interest in eugenics), casts human beings as inevitably and tragically bound by their "innate" qualities—a theme that resonates with the depictions of venereal disease throughout the rest of the novel and the eugenicist beliefs to which Grand subscribed.[45] Yet Angelica's story actually reflects a tragic reversal of this line of thinking, as she possesses inborn musical skills but lives in a world that does not let her fully realize them. Her "environment" indeed has quite a large "effect." Though, as Carol Senf argues, there is a hint of optimism in Angelica's assertion that "there will be plenty more like me by and by," for now, the vibrations of Angelica's violin simply cannot reverberate as strongly in the drawing room.[46]

"A Real Genius": Triumphant Musicality in *The Violin-Player*

While *The Heavenly Twins* ultimately relegates Angelica back to the domestic sphere, *The Violin-Player* launches Laurence into a wildly successful and permanent musical career. Thomas's novel represents perhaps the most triumphant narrative of female musicality in Victorian literature. Though Laurence must initially cross-dress to join Nielsen's studio, she ultimately enjoys a lasting musical life uninhibited by the trappings of marriage or domesticity.

Thomas was well situated to present such an optimistic narrative, as both of her sisters belonged to a rare cohort of women who enjoyed professional musical careers. Florence Ashton Marshall was a composer, conductor, and pianist; she attended the Royal Academy of Music, contributed to the first edition of Grove's *Dictionary of Music and Musicians*, wrote a biography of Handel, and conducted concerts for ensembles such as the South Hampstead Orchestra.[47] Thomas lived with Florence for a time and in fact wrote the libretto for Florence's operetta "Prince Sprite."[48]

Thomas's other sister, Frances, was a professional clarinetist, a career nearly unheard of for women during the Victorian period.[49] Frances studied with the British clarinetist Henry Lazarus at the Royal Academy of Music, where she won several awards for music performance and theory.[50] Though Lazarus generally found female clarinetists "unbecoming," he allowed Frances to play at his concerts and recitals.[51] After graduating from the Royal Academy, Frances performed widely throughout England at prestigious venues such as St. James's Hall and St. Martin's Hall and with ensembles such as the English Ladies' Orchestra, the Crystal Palace Ladies' Orchestra, and the Dundee Ladies' Orchestra.[52] In July 1887, the *Musical World* reported on Frances's contribution to a concert by the female violinist Gabrielle Vaillant:

> Even more interest was perhaps excited by Schumann's "Märchenerzählungen," a trio for piano, clarinet, and viola, in which the clarinet part was played by Miss Frances Thomas. Lady flautists are not now unknown, but it is indeed rare to find a lady undertaking the difficulties, combined as they are with the unpicturesque attitude, of the wind instrument in question. That she was quite equal to the occasion must be cordially admitted, her tone and execution being alike worthy of all commendation.[53]

Despite her society's condemnation of female musicality, Thomas lived alongside two women who disproved contemporary assertions that music was "beyond" them. Though a reviewer for *The Graphic* wrote that the events of *The Violin-Player* were "not altogether actual, or even possible," Thomas knew that a woman *could*, in fact, "find her true and whole life in art."[54]

In the opening pages of *The Violin-Player*, Thomas makes clear that Laurence (nicknamed Renza) is a born musical prodigy. In a section titled "The Little Rebel" in the novel's opening scene, a passerby named Mr. Romer wanders in the woods with his son and hears violin music—"a flashy, trashy air of Offenbach's"—emanating from a nearby enclosure.[55] The father and son stop to listen: "Certainly the precocious skill she displayed was sufficiently extraordinary, and even Mr. Romer must notice, with passing wonder, the nimbleness and flexibility of the little thing's fingers, the correctness of her intonation, and her spirited attempts at *bravura* passages on a toy-violin, probably picked up for a few francs at a fair."[56] A bystander remarks that Laurence must be a "prodigy," to which someone else in the crowd replies, "She must have been put to it in her cradle."[57]

The emphasis on Laurence's musical dexterity as a child resonates with Victorian-era scientific and pseudoscientific discussions of musical prodigies (like Gall's examination of Bianchi's musical skull) and the innateness of musical aptitude.[58] Laurence's musical aptitude—punctuated by her ability to play *bravura*, or virtuosic, passages—is rooted in her brain and body, present in her "nimble" and "flexible" fingers before she has even had any musical training (perhaps even while she was "in the cradle").

The organicism of Laurence's musical ability is further evident when readers learn that she possesses "perfect pitch" or "absolute pitch," the rare ability to hear notes in their natural key and replicate them. A few scenes after she plays her toy violin, Laurence comes across a house and is arrested by the sounds of music from within: "She listened entranced. Only a piano, a sweet voice, and a pretty valse. But it was long since Renza had heard anything so bewitching as that, anything better, indeed, than the scraping of her own violin-strings."[59] Laurence soon joins in; she "put[s] her fiddle in tune, and without a moment's hesitation beg[ins] to play over the air she had just heard from the garden, and that still haunted her head."[60] When one of the ladies inside the house asks her where she "pick[ed] up that tune," she replies, "Outside in the garden five minutes ago."[61] After one hearing, Laurence is immediately able to tune her instrument and replicate the melody she heard only a few minutes earlier.

Victorian music physiologists believed perfect pitch is something one is born with—a theory that endures today.[62] As the English scientist and music theorist R. H. M. Bosanquet (who also, incidentally, helped Alexander Ellis with revisions to his translation of Helmholtz's *Sensations*) wrote in his 1876 work *An Elementary Treatise on Musical Intervals and Temperament*,

> I think that most musicians will agree that those who have a very high development of the sense of absolute pitch have their ears altogether more finely strung, and more acute, than other people. That is to say if a man can tell me the exact sound of c and of any other note as he ordinarily uses them, without having any instrument to refer to, I consider that his musical organisation is such that his verdict on performances may be accepted without hesitation so far as their being in or out of tune according to his standard is concerned.[63]

Victorian scientists were especially fascinated by children who possessed the ability to name and replicate specific notes. An 1896 article in the *Musical Herald* discussed a case presented at the Royal Society of Edinburgh by the Glasgow physiologist J. G. McKendrick in which a four-year-old pianist named (aptly) John Baptist Toner displayed "the power of naming the absolute pitch of any sound."[64] Thomas makes clear that Laurence is one of these children with "finely strung" ears and a "musical organisation." Laurence's acute bodily proclivity for advanced musicianship is clearly not inhibited by her female gender.

However, Laurence's musical prowess does not initially help her infiltrate the male-dominated world of professional musicianship, since Nielsen rejects her from his studio. Yet, when Laurence returns to Professor Nielsen to audition dressed as a boy, she overwhelms him with her virtuosic skill: "Something took an abrupt hold of his attention. The *attaque* of the player in there had struck him. . . . Well, there was a purity of tone, a command of resources . . . most unusual in wonder-children."[65] Laurence exhibits unusual musical ability, power, and control. In particular, Nielsen's praise of Laurence's "purity of tone" echoes a compliment often paid to nineteenth-century violin virtuosos. In his 1836 violin manual, for instance, the music teacher George Dubourg attributes a "purity of tone" to legendary violinists like Paganini, Francesco Vaccari, and Carl Guhr.[66] In the context of Helmholtz's understanding that "pure tones" need to be "exactly regulated" to travel through the air with force and precision, Laurence's purity of tone further signals her sophisticated command over her playing.[67] She produces forceful, clear, articulate sounds that travel unhampered through space.

Laurence's playing not only manifests as the forceful actions of her own body but also affects Nielsen on a visceral level. As he listens to her play "a composition . . . full of difficulties—passages it might puzzle an advanced student to decipher correctly off-hand," his face displays "a contortion indicative of pleasure," and he "bend[s] forwards, listening with the intense and entire application of mind peculiar to those whose energies have all been appropriated to one purpose."[68] Through her music, Laurence is able to reverse the power dynamic established when Nielsen rejected her from the studio, gaining a form of physical control over his (male) body by causing his face to contort and his body to bend.

Laurence's powerful playing as the "boy" violinist incites physiological reactions in her friend and future love interest Damian Gervase as well. Midway through the novel, Laurence plays,

> soon becoming forgetful even of his [Gervase's] presence. At such moments the outside world was almost non-existent for the little musician. Laurence had early acquired the power of thus throwing all the nerve-force in her—so to speak—into one channel as she played. . . . "The little fellow's a real genius," thought Gervase, struck beyond all anticipation as he listened. He was the merest amateur in music; but genius is like the sun, its rays penetrate wise and ignorant alike. He was now contemplating the violin-player with an intentness of interest, which Laurence fortunately was too abstracted to perceive.[69]

Not only does the reference to Laurence's "nerve-force" further signal the embodied nature of her musicality, but the passage also shows the intense, palpable effects of her music on Gervase. He is "struck" by the penetrating power of the "boy" violinist's genius, "intent[ly]" focused on the music. Yet the scene also goes beyond a mere drawing-room courtship narrative. Laurence is too focused on her music to pay Gervase much attention; she becomes "forgetful of his presence" altogether.

The Violin-Player ultimately rewards rather than punishes Laurence for her gender subversion. When Laurence eventually reveals her identity to Professor Nielsen two years later, he agrees to continue teaching her and to help her launch her musical career. Nielsen insists, "If I forgive you, it is not because I take back my words, but because there is something higher and stronger than my will or than yours. Music is above even its professors."[70] Here, Nielsen hints that music can transcend the kinds of societal strictures that would limit Laurence's musicality based on her female identity. Gender, the novel asserts, does not determine musical skill—only the perception of it.

While Angelica's musicality withers once she sheds her disguise, Laurence's musical power only augments when she begins to perform as a woman. After Laurence reveals her identity, she plays "snatches of music" for a small audience: "Madame murmured delight; and Gervase was silent, all but his eyes. He made all sorts of excuses to himself for letting them utter the feeling that was insensibly mastering him."[71] Laurence channels the power to "master" her audience members—both male and female. After the performance, Gervase exclaims, "The miracles of electricity go no further!"—a phrase that punctuates the physical power of Laurence's music (and further tinges it with scientific resonance).[72]

As the novel progresses, readers glimpse Laurence's capacity to arouse physiological sensations in a wide variety of listeners. In one scene, her performance causes the face of the violin master Araciel to "bea[m] with pleasure and excitement."[73] In a later moment, her old friend Linda and her brother Bruno hear Laurence's music while they are traveling in Italy: "[Linda] stopped short, as a penetrating sound broke on the evening stillness. . . . The pair listened motionless awhile, holding their breath."[74] Even when she performs as a woman, Laurence ignites physical rapture in her audiences; her music penetrates the air and takes hold of her listeners' breath.

Laurence's musical influence is not only universal but also automatic and inevitable; her listeners cannot resist being affected by her playing. When Laurence performs at the opening of the Fenice Theatre in Venice, for example, she entrances her audience members with Hungarian airs that Nielsen worries will be "too fantastic" to appeal to the Italian listeners:

> That Laurence should make that music please those men was, in a technical sense, impossible. But frown and shrug though they might, they soon found they must look and listen too. As one after another of those wild stirring airs followed, the veriest Gallios present that night found the music they depreciated affect them strangely, keenly, delightfully. . . . To come down from the clouds, those would-be detractors were men, with heads, hearts, pulses, feelings, intelligence; and genius has a pass-key to all natures.[75]

Laurence's performance sends the audience into raptures, inciting "vociferous calls and recalls and deafening acclamations."[76] The power she exerts is tangible and visceral; it affects the "heads, hearts, [and] pulses" of her listeners—evoking physiological discussions of music's ties to the brain, muscles, and nerves. It is this corporeal power that renders her music a "pass-key to *all* natures" (emphasis mine).

Elsewhere, Laurence's violin playing excites an even more unlikely audience: the rats in her Bleiburg apartment. The narrator writes, "Laurence was wont to stay practicing almost without break from morning to night . . . fiddling away to an audience of rats, who listened spellbound or stopped their ears as might be, during the patient iteration of scales and exercises."[77] While this scene may appear silly or fanciful, it in fact echoes contemporary scientific discoveries of music's capacity to affect animals as well as humans, such as acoustical theorists' discoveries that animals possess hearing mechanisms that are just as—if not more—advanced as those of humans. The aurist David Tod, for instance, wrote, "The powers of music are very remarkable over a variety of animals. Only witness how fond the lizard is on hearing a lively air,—how he erects his head with one side generally higher than the other, and opens his mouth."[78] Similarly, the chemist and physicist William Hyde Wollaston argued that many animals and insects possess a wider range of hearing than do humans. Grylli insects, for example, display "the faculty of hearing still sharper sounds, which we do not know to exist."[79] In this context, readers might take literally Laurence's "spellbinding" influence over the rats; her music has the actual power to "stop their ears." Whether misogynistic music professors, traditionalist Venetians, or scurrying rats, Laurence's listeners cannot help but be enchanted by her music.

As with Valérie in chapter 2, Laurence performs so frequently and with such fervent bodily involvement that others fear her musicianship will result in her physical collapse.[80] In one scene, Laurence plays with such vigor that her listeners worry she is "draining the very springs of nervous energy."[81] Some, including her manager Emanuel Cuscus, fear she lacks the physical strength to keep performing. Cuscus insists that Laurence "puts too much of herself into her playing. . . . Women always do. They and their music are merged, not merely connected. If they had our strength, they would surpass us every way; but they haven't and their method wears them out."[82] These comments echo the sentiments of music teachers like Honeyman, who believed music was "beyond [women's] muscular power," as well as those who worried that musical performance would debilitate women's bodies. In his biography of the violinist Camilla Urso, the critic George Upton speculated that she might have dropped out of the musical world because her "physical strength had begun to wane."[83] Laurence's audiences become perversely fascinated by her dangerously energetic performances, which they believe might precipitate her demise: "People said she was killing herself, and flocked to hear her with redoubled alacrity."[84] Yet, in defiance of her naysayers, Laurence never falters; if anything, she gathers more strength and authority as her career continues: "To-day, as if to challenge and put to rebuke the last sceptics or detractors, she had selected to play some of the most trying pieces in a violinist's *repertoire*,—compositions certain to tax the finest faculties, and lay bare any weak point."[85] While the chorus of doubt surrounding her warns of the potential dangers of a musicianship so rooted in the body, Laurence's determination to redouble her efforts and showcase her strength serves as a powerful rejoinder to Victorian thinkers who dismissed women musicians for their lack of "muscular power."

In a drastic departure from other Victorian narratives about female musicality, including *The Heavenly Twins*, *The Violin-Player* ends with a triumphant celebration of Laurence's enduring performance career: "In Italy, Germany—all over the Continent, indeed—Mdlle. Therval now rejoices in a name so illustrious that he who should forget himself so far as to speak of her with moderation is likely to be branded as a raging iconoclast."[86] Though for most of the novel, Laurence resolves not to marry—insisting, "I will not forsake what I have lived for all these years"—she does ultimately fall in love with and consent to marry Gervase.[87] Yet, as with Valérie in chapter 2, Laurence's marriage further propels her musical career; Laurence feels she "touche[s] the tidemark of her highest force" when Gervase is in the audience.[88] After they become engaged, she embarks on a London tour and feels anew the "excitement of performing" and the return of her "old nervous energy."[89] Even after Gervase unexpectedly dies, devastating Laurence, she finds solace in performing and in fact relies on her music to survive. She "wanders through the world again alone, with a loyal old comrade—her violin."[90] Toward the end of the novel, her friend Val notes that Laurence's musical ability only seems to augment over time: "As she herself could never have played formerly: it was more forcible, earnest, and pathetic. To Val she seemed to have added something to the divinity of music by her genius for its interpretation. The clamorous applause in the theatre jarred on him."[91] Laurence's narrative thus ends not with marriage, domesticity, or death but with jarring applause.

The Violin-Player exposes the constancy of Laurence's musical power despite her shifting gender identities and life circumstances. Though Laurence is both a "she" and a "he" at different points, and the identities of her audience members (Nielsen, Gervase, Val, Englishmen, Hungarians, and even rats) also vary, Laurence's embodied musical life remains stable. Thomas creates a character whose musicality is so innate that, if given the opportunity, it transcends the social boundaries that would otherwise inhibit it. At the same time that the novel shows the body's manipulability and contingency through gender performance, then, it also makes space for a stable—and deeply pleasurable—corporeal life. In the world of *The Violin-Player*, there *is* a "doer over the deed"—a durable, musical body that persists despite "stylized acts."[92]

"Without Regard to Sex"

In a letter to the editor of the *Musical News* in 1893, a "Lady Student of Strings" wrote in favor of the widespread admission of women violinists to mainstream symphony orchestras: "Sex should not dominate in art. We should all have the same chances of success. . . . Such an end is surely not difficult to attain when one and all are fired with the enthusiasm which is born with the true love of the ideal."[93] Similarly, Florence Fidler, the advocate for women musicians mentioned in chapter 2, argued in 1900 that the "ideal orchestra is one consisting of men *and* women, each player being chosen without regard to sex, but for ability and conscientious discharge of duty."[94]

Whereas some nineteenth- and early twentieth-century musician-activists, most notably the members of the Society for Women Musicians, were committed to advocating for specifically *female* artists, both Fidler and the "Lady Student" decenter sex in their arguments in favor of a more universal approach that unites musicians "one and all," "without regard to sex." This "musician first, woman second" approach can certainly be read as an uncomfortable rejection of proto-feminist solidarity in favor of a kind of "equal opportunity," "gender-blind" ethic or assimilationist approach. Yet Fidler and the "Lady Student" also leave room for the possibility of an art form indeed *not* dominated by sex—of music as a pursuit that surpasses the restrictions imposed by gendered or sexual codes.

Along these lines, both Grand and Thomas present narratives of women who are, quite simply, not that interested in *being* women. Angelica and Laurence are musicians first, women second (and, in many key scenes, not women at all). Perhaps the most radical gendered implications of *The Heavenly Twins* and *The Violin-Player* are, ironically, that some things might be more important than gender.

Part Three
Sexualities

Chapter Four

Dangerous Vibrations

Musical Rape in George Eliot and Thomas Hardy

> It is at such moments as these, when a sensitive nature writhes under the conception that its most cherished emotions have been treated with contumely, that the sphere-descended Maid, Music, friend of Pleasure at other times, becomes a positive enemy—racking, bewildering, unrelenting.
>
> —Thomas Hardy, *Desperate Remedies* (1871)

Early in George Eliot's *Romola* (1863), the narrator describes an experience of sonic torture. As the painter Piero di Cosimo strolls through Florence's Piazza della Signoria, he cannot escape the pealing of the bells: "Piero di Cosimo was raising a laugh among them by his grimaces and anathemas at the noise of the bells, against which no kind of ear-stuffing was a sufficient barricade, since the more he stuffed his ears the more he felt the vibration of his skull."[1] Piero's body is at the mercy of the bells; he is physically overcome by their "unrelenting . . . vibration." His efforts to dampen the noise are not only futile but counterproductive; his ear-stuffing amplifies the sound, giving it a greater surface area on which to resonate. Piero sympathetically vibrates to the music he hears, but it is a vibration he does not want and cannot escape.

Most of this book centers on reparative readings of music's potential to produce gratifying pleasures and facilitate fulfilling forms of subjectivity, intimacy, and eroticism. While Ward, Francis, Grand, and Thomas all use acoustical science to imagine other, better possibilities for their characters' embodied lives, this chapter examines what happens when music precipitates harmful and unwanted physical sensations and violent social relations. Musicologist William Cheng describes the physical traumas of music torture: "For especially when music is extremely loud, repetitive, and imposed, it can do far more than touch. It pricks the skin, pummels the bone, penetrates the viscera, and unhinges the mind. It can discombobulate, traumatize, and humiliate. It breaks down subjectivity, rendering prisoners unable to hear themselves think. The vibrations, while invisible, do leave visible marks on their victims: twitches and tremors, the aftershocks of injury echoing in flesh."[2] Music is not always a source of bodily gratification; it can also be a tool for bodily ruin. As Cusick writes, "The thing we have revered for an ineffability to which we attribute moral and ethical value is revealed as morally and ethically neutral—as just another tool in human beings' blood-stained hands."[3] Aesthetic sensations can be complicit in violent acts.[4]

Nineteenth-century scientists and writers were aware of music's potential to harm. As discussed in earlier chapters, many acousticians understood music's physiological power as automatic, reflexive, and unavoidable. While a number of Victorian writers, such as those discussed in chapters 2 and 3, embraced the automaticity of musical response as a way to depict certain performers as "naturally" musical, others feared the implications of this inevitability: if humans' responses to music were automatic, weren't their bodies then vulnerable to forces—including violent ones—outside of their control?

In this chapter, I argue that both Eliot and Hardy—two writers who were especially musically inclined—depict insidious moments of musical influence in which human players weaponize music's physical effects to wield violent control over their listeners' bodies. Both authors envision the violent gendered and sexual dynamics that unfold when male players use their music to compel their female listeners to respond in automatic and uncontrollable ways. Focusing on Eliot's *The Mill on the Floss* (1860) and Hardy's *Desperate Remedies* (1871) and "The Fiddler of the Reels" (1893), I show that Eliot and Hardy depict male performers who use music to achieve physical contact with reluctant or resistant female listeners—a phenomenon I refer to as musical rape.

I use the term *rape* not to be provocative but to claim that these musical encounters *are*, indeed, actual instances of rape. The musical encounters I discuss are nonconsensual, harmful, and—in the context of acoustical science—deeply embodied; the male musicians produce particle-filled sound waves that enter the women's bodies through their ears and act upon their nerves and muscles. Some might claim that it is misguided to use the term *rape* here. After all, these musical encounters do not directly involve sexual organs (though, as discussed earlier, the association of sex strictly with genitalia is one feminist and queer theorists have long unsettled[5]). Nor would these moments have been called rape in the Victorian period. As Lana Dalley and Kellie Holzer note, rape—even when it was "clear"—was almost never recognized as such in Victorian England, particularly after the Offenses Against the Person Act (1861) left the definition of rape up to judges and juries, who almost always failed to prosecute due to the "normalization of irrepressible male sexuality, a contention that 'respectable' men were incapable of sexual assault, and an assertion that female victims of sexual assault were somehow to blame."[6] Elissa Gurman points out that nineteenth-century thinkers often described female sexual desire as something that, by definition, occurred without female agency; a desiring woman must be, as Havelock Ellis wrote, "swept out of herself and beyond the control of her own will, to drift idly in delicious submission to another and stronger will."[7]

In referring to these violent musical encounters as rapes, I draw on recent work by feminist rape theorists who argue for more "capacious" conceptions of rape that foreground the complexity, contradiction, and ambiguity that experiences of gendered and sexual violence so often involve.[8] Feminist philosopher Linda Martín Alcoff argues that theorists need a "new epistemology of rape" which refuses to think about rape as a "simple, straightforward matter."[9] For Martín Alcoff, it is futile to try to "establish the dividing line between harmful and harmless sex" or even to rely on the notion of consent, which is always already "embedded within structures that pose challenges for low-status groups of all sorts."[10] She suggests that activists and scholars focus on the range of "sexual violation[s]" that can happen in a variety of contexts: "With stealth, with manipulation, with soft words and a gentle touch to a child, or an employee, or anyone who is significantly vulnerable to the offices of others."[11] Sharing many of these critiques, Spampinato urges literary critics to move away from what she calls "adjudicative criticism," or the attempt to determine "whether rape has occurred" in a literary text or "whether fictional characters are guilty of rape."[12] Instead, Spampinato proposes a "capacious conception of rape" that "refuses to locate rape in a particular bodily act (as the law does), [that] refuses to yoke rape's harms to a particular gender, and [that] understands a variety of forms of violence as equally serious."[13]

For many feminist scholars, such "capacious conception[s]" of rape are useful because they bring to light a much broader range of bodily harms than are captured by traditional definitions of rape as penetrative or genital sex. Indeed, as Martín Alcoff notes, it was only in 2013 that the United States expanded the definition of rape to include "other orifices besides the vagina."[14] For literary scholars, defining rape capaciously means we might be able to take seriously *as* rape moments in texts that are not "triple-X explicit[]" but nonetheless reveal intense bodily and psychic harms.[15] Kimberly Cox, for instance, reads moments of "uninvited, unreciprocated" hand-grabbing in eighteenth-century novels not as "metaphor[s] for sexual violation" but as direct representations of "the act of sexual violence itself."[16] Similarly, Kathleen Lubey sees the "penetrative rape" in Samuel Richardson's *Clarissa* as only the " 'last' expression of the sexual incursions that have already been documented. . . . That momentary act only punctuates a rape that has already been in progress."[17]

It is in this context that I read several of Eliot's and Hardy's musical scenes as nothing less than instances of rape. This argument has multiple literary-critical and feminist-theoretical stakes. First, I show that music takes on much more than a figurative or symbolic role in depictions of sexual harm. Literary critics most often read scenes of male musical influence in nineteenth-century literature as either metaphorical (a stand-in for harms depicted off the page) or prophetic (a foreshadowing of dangers to come). Even more overt depictions of musical influence in Victorian literature, such as Svengali's mesmeric manipulation of *Trilby*'s titular heroine, are relegated to the supernatural realm, far from being "realistic" possibilities. Scholars of rape in Victorian literature also tend to focus on the many symbolic ways that Victorian writers captured rape. Most famously, Hardy's *Tess of the D'Urbervilles* (1891) has ignited over a century of debate over whether Alec did, in fact, rape Tess in the Chase and why Hardy renders the moment so "ambiguous."[18] Scholars have also identified intimations of gender-based violence in Eliot's works, such as the "clandestine" plot of "marital rape" that Doreen Thierauf brilliantly locates in *Daniel Deronda* (1876).[19]

Yet, in the context of nineteenth-century acoustical science, music presented a concrete, material mechanism for men to physically assault women's bodies, stimulating their nerves, muscles, and limbs in ways that were rarely consented to but, given the automaticity of musical sensation, impossible to avoid. Drawing on Taylor's framework of "aural sex," I argue that if rape is something that, like sex, can occur aurally as well as genitally, scenes of male musical influence emerge not as figurative nods to future harms but as potent moments of harm in and of themselves.[20] Defining rape capaciously allows readers to uncover moments of gendered and sexual violence that go far beyond vague or symbolic realms and to recognize Eliot's and Hardy's willingness to explicitly depict rape's phenomenological harms.

In some ways, then, I am participating in the very literary method that Spampinato critiques, as I claim that, in an acoustical context, rape *really does* occur in these works. However, my aim is not to "adjudicate" the "guilt[y]" fictional men, nor to examine their psychologies, but to highlight rape as a much broader, more diffuse, and more ubiquitous phenomenon—both in the nineteenth century and today—than scholars often deem it to be.[21] By showing how rape, capaciously defined, occurs even when there is no genital contact, even among other conflicting pleasures and desires, and even—or especially—in the context of irresistibly beautiful music that renders listeners submissive to its pleasures, I argue that Eliot and Hardy recognized the multiple, often unexpected ways that bodies—particularly poor, female ones—could experience harm.

This chapter reveals that Eliot and Hardy were attuned to many of the thorny, uncomfortable, and often devastating ideas about rape that feminist theorists still grapple with today. First, both Eliot and Hardy recognized that there is sometimes a deeply uncomfortable slippage—one frequently weaponized by misogynist culture in ways devastating to victims—between wanted pleasures and unwanted harms (what Martín Alcoff refers to as the blurry line between "harmful and harmless sex," or what Barbara Johnson calls the "undecidability between female pleasure and female violation").[22] Feminist theorists have long identified the troubling reality that rape sometimes elicits bodily responses that can be misread as signs of wanted arousal (vaginal wetness, penile erection, orgasm). As feminist legal scholar Catharine MacKinnon writes, "One particularly devastating and confusing consequence of sexual abuse for women's sexuality—and a crisis for consciousness—occurs when one's body experiences abuse as pleasurable. Feeling loved and aroused and comforted during incest, or orgasm during rape, are examples. Because body is widely regarded as access to unmediated truth in this culture, women feel betrayed by their bodies."[23] In the musical scenes here, some of the heroines initially desire the men's music or experience bodily pleasure in response to it, even when it is unwanted. Eliot and Hardy thus demonstrate what Martín Alcoff calls the "fluid and overlapping realities of our categories of coercive and non-coercive heterosexual sex."[24]

Eliot and Hardy also reveal the "complicated nature of culpability," especially in "male-dominant societies with epidemic amounts of gender-based forms of sexual violence."[25] Rarely, either in Victorian literature or in life, do we see clear depictions of genital violation with a defined perpetrator and victim (indeed, Spampinato outlines the folly of literary criticism that seeks only such depictions). It is often difficult to locate one single perpetrator of rape, as entire institutions, systems, and societies are often complicit in creating rape culture.[26] The presence of music in Eliot's and Hardy's works aids readers in deprioritizing narratives of sexual violence that center on one male perpetrator. In the musical scenes in this chapter, skin does not even touch; bodies are not even in close proximity; and there is not even one sole agent of rape, as sound waves collude with male players to perpetrate the assaults.

One of the greatest tragedies of this "complicated nature of culpability," Eliot and Hardy show, is that it makes it terrifyingly easy to blame victims and grant sympathy (or, to use Kate Manne's term, "himpathy") to perpetrators.[27] In the texts of these two authors, the presence of music blurs the role of the perpetrator; after all, it is not just the male *musicians* who rape but also their *music*, which, as acoustical science revealed, itself has the power to travel through space and enter women's bodies. If it is music that harms the woman, not the musician directly, is it *really* the perpetrator's fault? If the women are uniquely sensitive to music—as Eliot's and Hardy's heroines are—then aren't their own bodies partially to blame? As I will discuss, even twentieth- and twenty-first-century literary critics sometimes fall into this kind of "himpathetic" logic in their analyses of Victorian rape scenes, further obscuring the workings of sexual violence in both Victorian texts and contemporary culture.[28]

Finally, these musical rape scenes illustrate just how ubiquitous rape was in Victorian culture (and of course still is today). Music makes Eliot's and Hardy's female characters vulnerable to rape not only in bedrooms or wooded enclosures but also in drawing rooms, taverns, ballrooms, and concert halls—public spaces where their vulnerability and harm can be witnessed and weaponized by all those around them. While Eliot and Hardy do not always go as far to critique Victorian rape culture as modern-day readers might like—both authors were, after all, well known for their fatalistic worldviews, particularly when it comes to sexuality—they nonetheless offer important illustrations of how rape functions in Victorian society.[29] Like the vibrating sound waves that fill rooms and linger in the air, female bodily harm can be invisible—and yet all-encompassing.

"In Spite of Her Resistance": *The Mill on the Floss*

George Eliot's lifelong passion for music is well documented.[30] Eliot was a pianist and chamber musician, avid concertgoer, and acquaintance of famous performers like Franz Liszt, Anton Rubinstein, and Clara Schumann.[31] Her own work of music criticism, "Liszt, Wagner, and Weimar," appeared in *Fraser's Magazine* in 1855.[32] Eliot's musical philosophy most often reflected idealist or Romantic conceptions of music as a transcendent, sublime force and vehicle for moral good and human sympathy.[33] In poems such as "The Legend of Jubal" and "O May I Join the Choir Invisible!," as well as in her depictions of virtuous musical characters such as Dinah Morris, Daniel Deronda, and Caleb Garth, Eliot imagines music making and listening as "idealized form[s] of communal participation," as Nancy Henry writes.[34]

Yet, as Picker and da Sousa Correa outline, Eliot was also fascinated by physical and physiological acoustics.[35] Eliot and George Henry Lewes owned several editions of Helmholtz's *On the Sensations of Tone* and a collection of Helmholtz's popular lectures that included "The Physiological Causes of Harmony in Music" (1857).[36] Eliot's journals confirm her familiarity with Helmholtz's musical acoustics; in a February 1869 entry, she wrote, "I am reading about plants, and Helmholtz on music."[37]

Eliot was particularly interested in acoustical studies of musical influence, or the notion that music's intense physical and physiological effects gave the musician a great deal of power over the listener. For Eliot, musical influence was not always such a bad thing; in fact, she sometimes uses music as the productive force that shocks her characters out of states of stupor, sickness, or solipsism. In "Mr. Gilfil's Love Story" (1857), the "vibration" produced when Mr. Gilfil's nephew Ozzy strikes the piano awakens the heroine Caterina out of a long illness: "It seemed as if at that instant a new soul were entering into her, and filling her with a deeper, more significant life."[38] In Eliot's most famously musical novel, *Daniel Deronda* (1876), musical influence serves as an antidote to human solipsism and British bourgeois philistinism. For instance, when the Jewish virtuoso Julius Klesmer strikes a "thunderous chord" on the piano, its "sudden vibration" causes a panel near the piano to fly open and reveal the picture of a "dead face."[39] In response, the ever "sensitive" Gwendolen is sent into shock—a moment that, Picker writes, allows Klesmer to "puncture smug bourgeois facades" and gives Eliot the chance to, as she herself wrote, "widen the English vision a little in that direction [of Jewish people] and let in a little conscience and refinement."[40] Eliot seems to relish music's ability to literally shake humans out of their silliness, selfishness, and xenophobia.

However, Eliot puts musical influence to much more devastating use in *The Mill on the Floss* (1860), in which male musical influence acts as nothing less than a form of rape. Not only does Stephen Guest weaponize his vocal and piano music to manipulate and control Maggie's body, but a character named Torry also takes advantage of Maggie's musical pleasure to pressure her to dance with him. Music makes Maggie feel unwanted pleasures, causes her body to move in unusual and uncontrollable ways, renders her vulnerable to male advances, and causes her direct bodily harm.

Reading these musical encounters as moments of sonic rape reframes existing critical understandings of the novel's use of music and its gender and sexual dynamics, along with Eliot's broader engagement with sexual harm. Critics have long identified music's role in the novel as a symbol of Maggie's tragic fate(s): her social ostracization after her ill-fated boat ride with Stephen and her drowning in the flood. Weliver argues that Maggie's response to Stephen's music represents a "primitive emotional respons[e]" that ultimately leads her "into danger."[41] Similarly, da Sousa Correa finds an "almost Darwinian account of music" that "culminates in the opening of the floodgates of desire at the novel's end."[42] Both Picker and Burdett link the imagery of sound waves to the literal waves of water in which Maggie ultimately drowns at the end of the novel.[43] As Picker writes, the sound waves "aurally foreshadow . . . the surging river current that ultimately engulfs her."[44] Yet, in my reading, music does not simply prefigure Maggie's later traumas; music itself traumatizes her.

I propose a new reading of Stephen and Maggie's relationship and of the gender and sexual dynamics in the novel as a whole. Most critics interpret the fraught nature of Maggie and Stephen's encounters as a result of Maggie's moral obligations to Philip and Lucy, which are in conflict with her "real" or "true" sexual desires for Stephen. Gray writes, "Maggie is shocked—not from prudishness, but by the overt consequences of 'the sin of allowing a moment's happiness that was treachery to Lucy, to Philip—to her own better soul.' "[45] Weliver notes that Maggie "remembers Lucy's and Philip's feelings when she is not overwhelmed by Stephen's influence; her *conscious* sympathy for others means that she prioritizes her *conscience*."[46] But what if Maggie's encounters with Stephen represent fraught, traumatic experiences in and of themselves? Dalley and Holzer underscore the more troubling aspects of Stephen and Maggie's relationship, arguing that we can map their fateful boat ride onto contemporary understandings of "rape culture," as Maggie is "shamed and ostracized after an ill-conceived boat ride with her cousin's suitor who, despite publicly taking the blame for the incident, is left comparatively unscathed."[47] Vanessa Ryan also acknowledges that Maggie's "yielding" in the boat ride scene is "not a conscious choice and perhaps not a choice at all."[48] Indeed, the whole passage is filled with language of nonconsent: "without any act of her own will," "hardly conscious of having said or done anything decisive," "never consented to it with my whole mind," "deprive me of my choice."[49]

Reading Stephen's music making as a form of rape reveals the novel's engagement with sexual violence to be even more explicit and wide-ranging than critics have allowed. Music becomes a vector of gendered dominance that naturalizes Maggie's harm and exonerates Stephen. In the context of sonic rape, actual sexual harm comes to Maggie at several points throughout the text—not just during the boat ride but every time she hears Stephen's vibratory voice. While Eliot is rarely read as a particularly political writer, especially when it comes to gender and sexual politics, she was not at all unwilling to foreground the phenomenological horrors of gendered and sexual violence.[50]

Maggie, a lifelong lover of music, is at first enthralled by Stephen's singing. Though Stephen is a "provincial" and "amateur" singer whose voice, the narrator tells us, "would have left your critical ear much to desire," Stephen's music offers Maggie a thrilling alternative to her otherwise provincial life and relieves her from a long period of sensual deprivation (a period patently defined by a *lack* of music, with "no piano, no harmonised voices, no delicious stringed instruments").[51] The "vibratory influence" of Stephen's singing revives Maggie in ways that are at first described as intensely pleasurable and even rejuvenating to her body.[52] One night, after listening to Stephen sing, she goes upstairs to her room and feels the music "vibrating in her still."[53] She paces the floor "with a firm, regular, and rather rapid step, which showed the exercise was the instinctive vent of strong excitement," and displays "an almost feverish brilliancy" in her eyes and cheeks: "Her head was thrown backward and her hands were clasped with the palms outward and with that tension of the arms which is apt to accompany mental absorption."[54] As she tells her cousin Lucy Deane, "I think I should have no other mortal wants, if I could always have plenty of music. It seems to infuse strength into my limbs and ideas into my brain. Life seems to go on without effort, when I am filled with music. At other times one is conscious of carrying a weight."[55] Stephen's music invigorates Maggie and gives her a sense of as yet unexperienced vitality. However, Eliot is careful to distance the experience from Stephen himself; he is removed from the scene both physically, as Maggie is upstairs, and linguistically, as "music"—not

Stephen—is the subject of most of the sentences. Music itself, rather than any one person or thing, is Maggie's greatest "mortal want[]."

But even here, the narrator warns readers to be wary of all of this musical pleasure. In the middle of the scene, the narrator interjects that Maggie's musical susceptibility may prove a liability: "Such things could have had no perceptible effect on a thoroughly well-educated young lady with a perfectly balanced mind, who had had all the advantages of fortune, training and refined society."[56] "Poor Maggie," however, has a "highly strung, hungry nature."[57] On one hand, this moment can be read as the narrator blaming Maggie for her bodily vulnerability (and as I will discuss later, several literary critics do read the scene in this way). Yet the narrator attaches Maggie's vulnerability not to a personal failing but to her fatal class position (her "poor education and training"). (And in fact, the word *perceptible* leaves open the possibility that similar sensations might be occurring in more well-educated ladies; they are just not as perceptible. Might Lucy Deane's body also be buzzing and vibrating to the music, but she is just better trained to conceal it?) In line with Eliot's often-fatalistic treatment of gendered violence, Maggie's susceptibility to sexual harm is all but inevitable.

Indeed, Stephen soon weaponizes his "vibratory influence" against Maggie. A few scenes later, Stephen's music becomes a form of nonconsensual physical contact that Maggie does not want but is unable to resist:

> Maggie always tried in vain to go on with her work when music began. She tried harder than ever today, for the thought that Stephen knew how much she cared for his singing, was one that no longer roused a merely playful resistance, and she knew too that it was his habit always to stand so that he could look at her. But it was of no use; she soon threw her work down, and all her intentions were lost in the vague state of emotion produced by the inspiring duet—emotion that seemed to make her at once strong and weak, strong for all enjoyment, weak for all resistance. When the strain passed into the minor she half started from her seat with the sudden thrill of that change. Poor Maggie! She looked very beautiful when her soul was being played on in this way by the inexorable power of sound. You might have seen the slightest perceptible quivering through her whole frame, as she leaned a little forward, clasping her hands as if to steady herself, while her eyes dilated and brightened into that wide-open, childish expression of wondering delight which always came back in her happiest moments.[58]

In some ways, the narrator sanitizes the scene by emphasizing Maggie's beauty and "childish expression" and by tying the experience to the "delight" of her "happiest moments." There is even some ambivalence about how bad the moment *really* is; Maggie is "at once strong and weak; strong for all enjoyment, weak for all resistance."

And yet, just because Maggie experiences "enjoyment" here does not mean that this is not a scene of aural rape. Stephen wields undeniable physical control over Maggie's body, not only through his sly and deliberate positioning and gaze ("it was his habit always to stand so that he could look at her") but also through his music. The effects of his music are intensely physical; Stephen's shift into a minor key causes Maggie to "half start[] from her seat with [a] sudden thrill," renders her "at once strong and weak," and leaves her frame "quivering" and her eyes "dilated." These intense physiological effects are, crucially, unwanted. Maggie's "resistance" is "no longer . . . merely playful." She "trie[s] harder than ever" to return to her work, but her "intentions" are "lost." While readers could interpret her "quivering frame" and "clasped hands" as evidence of the ignition of sexual passion, it is crucial to remember that this is a passion that Maggie *does not want* to experience. Her music takes away her will ("intention") and even consciousness ("vague state of emotion"), rendering her "childlike" (a word that in my reading reflects not enjoyment but dependency). As Elisha Cohn argues, Stephen "acts on Maggie as a drug, rendering her unable to think beyond the moment. Her actions are all reactive—all supplemental outward stimulation and no inner response, such that sensation evacuates interiority."[59] In the context of contemporary musical science, Cohn's drug analogy becomes even more literal. Stephen alters Maggie's body *and* mind; his music is a physical substance that directly affects Maggie's physiological state. Even syntactically, this passage signals Maggie's lack of agency, as it shifts from passive to active voice: "She soon threw her work down, and all her intentions were lost." Moreover, language like "always," "habit," and "harder than ever" highlight the routine, consistent nature of the assault. Finally, it is both Stephen *and* his music ("the inexorable power of sound")—and, perhaps, not even just *his* music but Lucy's as well (this is, after all, a duet)—that deprive Maggie of her agency and autonomy; the threats of bodily violation are ubiquitous.

Stephen soon seizes another, even more overt, opportunity for musical assault. After Philip Wakem sings a lovely but passionless sentimental song (one that "touche[s] not thrill[s]" Maggie), Stephen takes sonic control of the space, making "all the air in the room alive with a new influence."[60] Maggie cannot help but be moved by Stephen's singing: "Maggie, in spite of her resistance to the spirit of the song and to the singer, was taken hold of and shaken by the invisible influence—was borne along by a wave too strong for her. But angrily resolved not to betray herself she seized her work, and went on making false stitches and pricking her fingers with much perseverance, not looking up or taking notice of what was going forward."[61] Here, both the music and the musician—the narrator merges "song" and "singer"—exert more overtly violent bodily control over Maggie, as the song takes hold of and shakes her. The music precipitates an instance of actual, injurious bodily penetration when she loses control over her fingers and punctures her skin with a needle. (The fact that this wound is partially self-inflicted makes this moment all the more cruel, suggesting that Maggie is complicit in her own violation.) Again, these events occur "in spite of her resistance" and despite the "perseverance" with which Maggie endeavors to keep to her work.

Stephen is not the only man who takes musical advantage of Maggie's body. At a dance late in the novel, a young man named Torry seizes on Maggie's responsiveness to music in order to convince her to dance with him:

> Maggie at first refused to dance, saying that she had forgotten all the figures—it was so many years since she had danced at school; and she was glad to have that excuse. . . . But at length the music wrought in her young limbs, and the longing came; even though it was the horrible young Torry who walked up a second time to try and persuade her. She warned him that she could not dance anything but a country dance, but he, of course, was willing to wait for that high felicity, meaning only to be complimentary when he assured her at several intervals that it was a "great bore" that she couldn't waltz—he would have liked so much to waltz with her. . . . She felt quite charitably towards young Torry, as his hand bore her along and held her up in the dance; her eyes and cheeks had that fire of young joy in them which will flame out if it can find the least breath to fan it.[62]

Maggie's response calls to mind work in physiological acoustics about humans' inevitable bodily vulnerability to dance music. Allen detailed humans' susceptibility to rhythm:

> The nervous system . . . put[s] itself into a position of expectancy, and is ready for the appropriate discharge at the right

moment. . . . The due recurrence of all such periodic actions is pleasant, because the organs perform their functions at the exact moment of expectation. . . . In dances, especially such as waltzes, polkas, and galps, where the rhythm is constant, we see the simplest conscious utilization of the pleasure arising from this measured recurrence. When once we have mastered the first difficulty of learning the step, the easy mechanical nature of the periodic motion gives us abundant muscular exercise in the form most nearly approaching the aesthetic enjoyments.[63]

Not following these bodily compulsions, or having them interrupted, Allen explains, can feel like an "unpleasant jar, and may entirely upset the harmonious actions of the limbs for a few minutes."[64] Similarly, in *How We Hear: A Treatise* (1901), music critic Frederick Charles Baker wrote somewhat deprecatingly of dance music, which acts upon "the motor and sensory nerves only, and may be truly said to serve the flesh more than the spirit. . . . It affects our heart pulsations, so that we feel excited and lighthearted, but from such music we never gain noble thoughts, or aspire to greater things."[65] In the dance scene, Maggie's nervous system drives her to relent to the "horrible" young man's persistent pressure to dance because she cannot avoid being kinesthetically affected by the music that envelops the ballroom. She experiences the stimulation of her "limbs" and the irresistible urge ("longing") to dance. Though the dance gives her a kind of joy (though a precarious one, as the "fire" in her eyes and cheeks can easily be fanned out with the "least breath"), and she even feels "charitably" toward Torry at one point, the music renders her powerless to refuse him and susceptible to an even more tactile moment of male bodily control, as Torry's hand "b[ears] her along and h[olds] her up." In this moment, Eliot envisions music as something that so affects Maggie's limbs that she must be propped up and moved across the dance floor against her will (despite her refusal). In line with Victorian sexual politics that rendered women complicit in their own victimization, Maggie's physiological sensitivity to music deprives her of sexual agency.

This particular dance further victimizes Maggie to Stephen and renders her an object of male sexual exchange. Stephen is overcome with violent jealousy as he watches Maggie dance with Torry: "His eyes were devouring Maggie: he felt inclined to kick young Torry out of the dance, and take his place. . . . The possibility that he too should dance with Maggie, and have her hand in his so long, was beginning to possess him like a thirst."⁶⁶ It is in the very next scene, in fact, that Stephen haptically violates Maggie; "seized" by a "mad impulse," he "dart[s] towards the arm, and shower[s] kisses on it, clasping the wrist."⁶⁷ Maggie's trauma is clear; in a "deeply shaken, half-smothered voice," she replies, "How dare you?" and, "quivering with rage and humiliation," "thr[ows] herself onto the sofa, panting and trembling."⁶⁸ However, Stephen's hand-grabbing is just one of the many bodily violations he has enacted throughout the text; while here he directly touches her skin, he had long been touching her nerves and muscles with his music.⁶⁹

Literary critics are often curiously quick to exonerate Stephen—and, in a way, to blame Maggie—for these vexed encounters. Gray argues that Maggie and Stephen enjoy a "kind of willing consummation" and that Stephen "is guilty of nothing worse" than a "kind of minor betrayal[] that most of us daily practise and daily forgive ourselves."⁷⁰ Weliver writes, "Stephen does influence Maggie through music, but it is not distinctly deliberate. . . . Although it may seem that Stephen deliberately manipulates Maggie like a mesmerist would, he actually struggles with self-mastery and loses."⁷¹ In such moments, "Maggie seems played upon but . . . a particular person is not at fault so much as the reaction of her unconscious mind to external stimulus."⁷² Though Rachel Ablow recognizes that Maggie's vulnerability to Stephen's influence is "extremely dangerous," she focuses more on Maggie's state of mind—her "powerless absorption in Stephen" and her "struggl[e] to remain aloof and unaffected in Stephen's presence"—than on Stephen as a perpetrator.⁷³ It is understandable that critics would hesitate to find Stephen "guilty" of anything more than a "kind of" vague, unintentional influence (and, as Spampinato would likely note, it is problematic to even be "adjudicating" Stephen in the first place). After all, Maggie at times experiences undeniable—and at times desired—pleasure in response to Stephen's music. Moreover, the narrator repeatedly blurs the music and the musician (the "singer" and the "song") in a way that might seem to absolve the men (Stephen or Torry, who takes advantage of the music already playing). And yet, if readers interpret these moments as nothing less than musical rapes, then attributing the violations to Stephen's loss of "self-mastery" or to Maggie's own susceptibility risks perpetuating patriarchal impulses of himpathy or victim blaming that, historically, have been constitutive of rape culture. This is perhaps what makes Eliot's depictions of musical rape so profound—and so profoundly horrifying: she underscores some of the deepest complexities, ambiguities, and contradictions in both Victorian and modern-day discourses about rape. Sometimes rape occurs after or coincides with moments of otherwise ambivalent desire or pleasure (such as Maggie's various experiences of "longing," "thrill," and even "wondering delight"). Sometimes rape can be caused by people who elsewhere show kindness (as Stephen at times does) or who do not necessarily intend to rape.[74] Sometimes it is painfully easy to blame the victim and exonerate the perpetrator, as even modern-day literary critics do. And sometimes rape is caused not just by one person but by a whole context or culture that facilitates and perpetuates it, as it is Stephen himself and also his music, Torry, and the music of others that collude to victimize Maggie. Perhaps Eliot's most powerful intervention, then, is that she highlights the enduring impossibilities of fully recognizing, naming, and pinpointing the exact contours of rape while still demonstrating its undeniable harms. Rape is both perpetuated by individuals *and* unevenly dispersed across systemic fields of power and privilege.

"Excruciating Spasms": *Desperate Remedies* and "The Fiddler of the Reels"

Hardy was "rapturous about music," particularly the folk music that evoked his family history and the landscape of Dorset.[75] His grandfather, father, uncle, brother, and cousins were all parish musicians, and Hardy himself learned the fiddle at a young age and played at village dances and revels throughout his life.[76] Hardy was also interested in classical music; he read George Grove's *Dictionary of Music and Musicians* (1878), attended performances by Tchaikovsky and Wagner, and corresponded with the composer Gustav Holst.[77] In August 1927, Holst even asked Hardy if he could "dedicate [his] new orchestral composition" to him "as a little token of respect and gratitude. It is entitled 'Egdon Heath' and is the result of reading the first chapter of 'The Return of the Native' over and over again and also of walking over the country you describe so wonderfully."[78]

Hardy's passion for music was scientific as well as personal. Like Eliot, he was fascinated by contemporary discoveries in musical science. He had strong opinions about the ongoing debate among Spencer, Darwin, and Gurney about whether music originated from human speech or vice versa.[79] In the margin of one of his notebooks, Hardy wrote, "MR. S. is right."[80] At the same time, Hardy was becoming increasingly interested in physiological science. He regularly corresponded with Allen and studied the works of other leading physiologists like Lewes and T. H. Huxley.[81] Hardy copied the following statement from Lewes into one of his notebooks: "Physiology began to disclose that all the mental processes were (mathematically speaking) functions of physical processes, i.e.—varying with the variations of bodily states; & this was declared enough to banish for ever the conception of a Soul."[82] Hardy was thus preoccupied by the notion that humans' bodies determine much of their being—an insight that critics such as Morgan, Elaine Scarry, John Hughes, and Suzanne Keen have traced in his fiction.[83] For instance, Hardy wrote in his notes, "Mr. M. Arnold . . . insisted somewhat too strenuously on the purely intellectual & moral aspects of art. There is a widely different way of regarding the same subject matter, which . . . dwells upon sensuous presentation, emotional suggest^n, & technical perfection as the central & essent^l qualities of art."[84] Judging from Hardy's journals, Spencer's work helped him tie many of these new physiological principles to the study of musical sensation. Hardy owned Spencer's *Essays: Scientific, Political, and Speculative* (1868–78), the second volume of which contains "On the Origin and Function of Music"—the essay that linked musical sensation to muscular excitement.[85] He copied the following line from Spencer's *The Principles of Biology* (1864) into his notebook: "The Deaf Dr. Kitto described himself as having become exceedingly sensitive to vibrations propagated through the body; & as so having gained the power of perceiving as by ears"—a passage he would later reference in *The Return of the Native* (1878) when he describes Eustachia using her ears to navigate the heath.[86] Hardy also read, annotated, and took notes on works such as Gurney's "On Some Disputed Points in Music" (1876), which refers at length to the theories of Spencer and Helmholtz, and George Romanes's "World as an Eject" (1886), which discusses Helmholtz's ideas of sympathetic vibration.[87] Hardy's note on Romanes, with his annotations in parentheses, reads, "The objective explan [sic] (scientific) given by Helmholtz of the effects of a sonata on the human brain (e.g., number of vibrations, &c.)."[88] Hardy was thus captivated by the intricacies of sonic vibration and the embodied nature of musical response.

Scholars have written at length about the role of music in Hardy's life and works. Critics often trace the references to music (especially folk music) in Hardy's poetry and prose as reflections of his nostalgia for his homeland, his engagement with the Dorset landscape, his celebration of rural working-class traditions, his interest in memory, and his anxieties about the encroachment of modernity on folk life.[89] Hughes and Mark Asquith have discussed Hardy's use of music, including his depictions of musical influence, as tied to his deterministic (and often tragic) worldview.[90] Asquith writes, "[Hardy's] tragic characters are simply out of tune with the world that surrounds them."[91]

Hardy also depicts music as a tool of sexual assault. In *Desperate Remedies* and "The Fiddler of the Reels," Hardy puts embodied musical influence in the hands of two of his most despicable sexual predators: Aeneas Manston and Mop Ollamoor, both of whom wield their musical skills to physically harm women. My reading of musical rape in Hardy sheds new light on his engagement with gendered and sexual violence. Hardy is the Victorian author perhaps most associated with rape, given *Tess of the D'Urbervilles*'s (in)famous (and often censored) Chase scene. Literary critics have debated the "rape versus seduction" question in *Tess* for over a century.[92] Yet the musical traumas enacted by Aeneas and Mop expand existing critical discussions of rape in Hardy. Aeneas and Mop physically violate their listeners' bodies with their "excruciating," "gnawing," injurious, and, most crucially, *unwanted* music. In this context, we can see that Hardy depicted rape not only in vague references to "gossamer" and "guardian angel[s]," as he does in *Tess*, but also in more overt descriptions of musical bodies that vibrate without their control or consent.[93]

The plot of *Desperate Remedies* surrounds the victimization of the heroine Cytherea Graye. Cytherea suffers at the hands of familial, social, and economic forces from the beginning of the novel. She is left destitute by her father, who dies without repaying his debt; obligated to provide for her brother when he hurts his leg; and jilted by her lover Edward Springrove, who is betrothed to another woman without her knowledge. These circumstances thrust her into the orbit of the adulterous and murderous Aeneas, whom readers eventually learn is the illegitimate son of Cytherea's employer, Miss Aldclyffe. Aeneas fakes the death of his first wife, then actually kills her, then convinces another woman to take her place, then poisons *her*, then finally tries to kidnap Cytherea before ultimately confessing and killing himself in jail. He is also, crucially, a musician. Far from merely symbolizing his despicable character, however, Aeneas's musicality actually serves as a tool in his violation of Cytherea.

Hardy makes it clear early in the novel that music will be a source of harm for Cytherea. She, like Maggie, is cast as uniquely susceptible to music. Throughout the novel, Cytherea is attuned to distressing noises and sounds, such as those that disturb her at night in Miss Aldclyffe's house. On her first night there, she is tortured by the music of a well: "Now that she had once noticed the sound there was no sealing her ears to it. She could not help timing its creaks, and putting on a dread expectancy just before the end of each half minute that brought them."[94] Cytherea's misophonic anxieties heighten when another sound—"a very soft gurgle or rattle"—acts "so powerfully upon her excited nerves, that she jumped up in the bed."[95] Elsewhere, the rattling of trees in the wind lead her to dream that she is "being whipped with dry bones suspended on strings, which rattled at every blow like those of a malefactor on a gibbet; that she shifted and shrank and avoided every blow, and they fell then upon the wall to which she was tied."[96]

Cytherea's sonic nightmare comes true when she hears Aeneas play. One day, as Cytherea is walking back to Miss Aldclyffe's, she perceives the approach of a thunderstorm and makes her way toward an old manor house. She encounters Aeneas, the steward of the manor, on the steps and immediately feels "his eyes . . . going through [her]."[97] Cytherea's encounter with Aeneas is rife with intimations of violence and nonconsent from the beginning. He teases her, "Are you afraid? . . . Of the thunder, I mean . . . not of myself."[98] He then asks her if he can walk up to the house with her, and though she refuses, he eventually convinces her to approach the house by promising to fetch a subscription check that he owes Miss Aldclyffe. Soon, the weather "compel[s] her, willingly or no, to accept his initiation."[99] Aeneas uses the rain as justification to lure Cytherea all the way inside the house—with only the vague promise that "an old woman [is] . . . in the back quarters somewhere."[100]

Once they are inside, Aeneas leads Cytherea to his homemade organ, the "only piece of ornamental furniture yet unpacked."[101] As the thunderstorm "shake[s] the aged house from foundations to chimney," Aeneas approaches the instrument, and Cytherea "uneasily" asks, "You are not going to play now, are you?"[102] Aeneas replies that he will and—"without waiting to see whether she sat down or not"—begins "extemporising a harmony which meandered through every variety of expression of which the instrument was capable."[103] As he plays, "running his fingers again over the keys," he asks her to look at him, and she feels "compelled to do as she was bidden."[104] When Aeneas asks if she wants to hear another piece, she says no, but he "commence[s] without heeding her answer, and she stood motionless again, marvelling at the wonderful indifference to all external circumstance which was now evinced by his complete absorption in the music before him."[105] Aeneas not only curates a decidedly Gothic moment—the storm, the old manor, the music—that "frighten[s]" Cytherea with its "unearthly weirdness," but he also deliberately ignores both her body language and her (lack of) consent—so much so that Cytherea marvels at his "indifference."[106] As he physically manipulates the instrument ("running his fingers again over the keys"), he also gains control over her body, compelling her to look in his direction.

As the scene continues, Aeneas's music becomes an even more forceful agent of bodily manipulation:

> He now played more powerfully. Cytherea had never heard music in the completeness of full orchestral power, and the tones of the organ, which reverberated with considerable effect in the comparatively small space of the room, heightened by the elemental strife of light and sound outside, moved her to a degree out of proportion to the actual power of the mere notes, practised as was the hand that produced them. The varying strains—now loud, now soft; simple, complicated, weird, touching, grand, boisterous, subdued; each phase distinct, yet modulating into the next with a graceful and easy flow—shook and bent her to themselves, as a gushing brook shakes and bends a shadow cast across its surface.[107]

Hardy uses the language of acoustics to highlight the intensity of Aeneas's influence; his music is too loud for the small space and thus "reverberate[s]" violently, taking power over both the atmosphere and Cytherea's body. Aeneas uses musical variation (produced by his "practised . . . hand") to manipulate Cytherea's body, to literally "shake" and "bend" and "sway" her to his will. In the "gushing brook" analogy (no doubt a nod to her manipulation by sound waves), Cytherea is not even a full presence but a "shadow" borne along by the current.

The narrator then becomes even more explicit about the music's power over Cytherea's body. It "shift[s] her deeds and intentions from the hands of her judgment, and hold[s] them in its own. . . . She was swayed into emotional opinions concerning the strange man before her; new impulses of thought came with new harmonies, and entered into her with a gnawing thrill."[108] This scene is not without pleasure, and yet it is a violent (and of course unwanted) pleasure; the phrase "entered into her with a gnawing thrill" casts this moment as one of actual bodily penetration. As in *The Mill on the Floss*, the narrator shifts the agency from Aeneas himself to the music (the "power of the music" and "new harmonies"). While the fact that it is music (rather than another of Aeneas's bodily "organs") might seem to absolve him of the rape, I argue that this detail actually works to heighten the trauma of the scene. Aeneas's assault of Cytherea is no less bodily because it occurs through music, and it is just as (if not more) overwhelming; the music that assaults her surrounds and envelops her, vibrating through the air and acting upon her whole body without her control.

Soon Aeneas's musical influence draws Cytherea physically closer to him: "She found herself involuntarily shrinking up beside him, and looking with parted lips at his face. . . . She was in the state in which woman's instinct to conceal has lost its power over her impulse to tell; and he saw it."[109] Aeneas "bend[s] his handsome face over her till his lips almost touched her ear" and murmurs to her that he could "see [she was] affected by it."[110] The music has facilitated his further physical encroachment on her body—not incidentally on her ears. Like Stephen, Aeneas is aware of his music's power to affect his victim, and like Maggie, Cytherea is evacuated of her ability to resist and "involuntarily" compelled to approach him.

Hardy eventually allows Cytherea to shed the effects of Aeneas's musical manipulation. When the music ends, his "influence over her . . . vanishe[s] with the musical chords," and she is able to "tur[n] her back upon him."[111] And yet, the harm lingers. Cytherea realizes "she would have given much to be able to annihilate the ascendency he had obtained over her during that extraordinary interval of melodious sound."[112] She reflects, "She was full of a distressing sense that her detention in the old manor-house, and the acquaintanceship it had set on foot, was not a thing she wished. It was such a foolish thing to have been excited and dragged into frankness by the wiles of a stranger."[113] While it is still a long time before Cytherea fully escapes Aeneas's clutches, she is not entirely doomed like Maggie (or, as I will show, Hardy's other heroine Car'line Aspent). Yet *Desperate Remedies* still issues a potent reminder of just how vulnerable Victorian women—especially young, poor ones wandering alone in bad weather—are to rape.

An even more explicit musical rape plotline unfolds in "The Fiddler of the Reels," a short story published two years after *Tess*. The antihero, a fiddler named Wat Ollamoor (nicknamed "Mop" because of his long hair), is known for his "weird and wizardly . . . power over unsophisticated maidenhood."[114] Though Mop is coded as hypersexual and foreign (a "woman's man," a "dandy," and "rather un-English"), Hardy is careful not to attribute his sexual predation solely to his otherness.[115] Mop's influence over "unsophisticated maidenhood" is envied by "many a worthy villager"; the desire for control over female bodies is universal among all men, whether English or not.[116] Though Mop is not a particularly skilled musician due to his "indolence and averseness to systematic application," there are "tones in [his music]" that suggest his potential to become a "second Paganini"—a reference that connotes potentially dangerous musical influence, as widespread rumors that the Italian violinist Niccolò Paganini was a "demon" circulated throughout nineteenth-century musical culture.[117]

Mop's music entices everyone around him. In one moment, he appears as a kind of perverse Pied Piper character who can "make any child in the parish, who was at all sensitive to music, burst into tears in a few minutes by simply fiddling one of the old dance-tunes he almost entirely affected."[118] Mop maniacally wields his musical power, "laughing as the tears rolled down the cheeks of the little children hanging around him."[119] His musical influence is so strong, in fact, that the narrator imagines that it might even be able to affect physical objects, drawing "an ache from the heart of a gate-post" or "tears from a statue."[120]

Mop's physically arousing music has the strongest effects on women, "especially young women of fragile and responsive organization."[121] The tragic heroine Car'line is one such woman. As the narrator tells us at the beginning of the story, Mop's "heart-stealing melodies" bring Car'line "discomfort, nay, positive pain and ultimate injury."[122] Like Maggie and Cytherea, Car'line is uniquely sensitive to sound and indeed to all kinds of bodily stimulation; she has a particularly "responsive organization," and her father fears that her "hysterical tendencies" and "spasmodic little frame" will result in a "species of epileptic fit."[123]

Car'line first encounters Mop's music one evening when she is walking through Lower Mellstock. Mop stands in his doorway, playing the fiddle, the music of which ignites in Car'line a "wild desire to glide airily into the mazes of an infinite dance."[124] Though Car'line tries to keep walking, "her tread convulse[s] itself more and more accordantly with the time of the melody, till she very nearly dance[s] along. . . . Her gait could not divest itself of its compelled capers till she had gone a long way past the house; and Car'line was unable to shake off the strange infatuation for hours."[125] In acoustical terms, Car'line's body is in uneasy vibratory sympathy with Mop's music. Her movements—explained by contemporary scientific discoveries of music's kinesthetic effects (such as Allen's comments about dance music and Gurney's understanding of the "direct impulse to move" that melody can ignite)—are clearly unwanted ("could not divest itself," "compelled capers"); she appears as a puppet responding automatically to another's influence.[126] In these moments, Car'line's body is not her own; even linguistically, the double "itself" suggest her lack of conscious involvement, as her body moves on its own, without her will. The narrator makes clear that Car'line's reaction is unavoidable, a biological phenomenon rooted in her nervous system; readers learn that it would "require a neurologist" to explain Mop's influence over Car'line.[127]

At first, Mop's music intrigues Car'line and brings her pleasure. In fact, she "contrive[s] to be present" anytime Mop is playing.[128] The narrator contrasts Car'line's lukewarm response to her suitor Ned Hipcroft with the erotic arousal she experiences when Mop is playing: Ned has "not the slightest ear for music" and cannot "play the fiddle so as to draw your soul out of your body like a spider's thread, as Mop did, till you felt limp as a withywind and yearned for something to cling to."[129]

Yet Mop's musical power over Car'line soon leads directly to at least one moment of off-screen carnal contact that threatens to ruin her. A few scenes after the initial musical encounter, readers learn that Car'line has become pregnant and that Mop has abandoned her, leading her to beg Ned to take her (and her illegitimate child) back. She tells Ned in an all but outright acknowledgment of the rape (and in language strikingly reminiscent of Tess Durbeyfield's lamentation that her mother did not tell her "there was danger in men-folk"), "And I never had a young man before! And I was so onlucky to be catched the first time he took advantage o'me."[130] Mop's music not only provides a vague foreshadowing of literal sex but is itself part of his sexual manipulation, rendering her "limp as a withywind" and inducing in her a range of sensations that she is unable to shake off "for hours." Mop's musical and sexual rape of Car'line physically harms her in the moment and also results in continued dire consequences for her body. The "positive pain and ultimate injury" that the narrator warned of were very real. As Spampinato writes, "Sex was *literally* and socially dangerous for nineteenth-century women; maternal mortality was high, and pregnancy outside of marriage spelled economic destitution as well as social erasure. The threat of rape was particularly damning because with it came all the consequences of consensual sex as well as the likely erasure of the victim as a legal subject, since allegations of rape were rarely prosecuted."[131] While readers might think Hardy only hints at a sexual encounter that occurs outside of the text, he actually presents a rather vivid depiction of the literal violation that occurs through Mop's fiddle—one that has disastrous consequences for Car'line's body, safety, and social position.

As the story progresses, Mop continues to musically rape Car'line. Later, when Ned and Car'line travel back to South Wessex from London, Ned gets off the train in Casterbridge to make employment inquiries, and Car'line and her daughter walk to Stickleford. They stop at a hostel to rest, where Car'line is immediately plied with copious amounts of "gin-and-beer" ("though she did not exactly want this beverage")—already being stripped of her bodily agency before she even hears a note.[132] Lo and behold, she sees Mop in the corner rosining his bow in preparation for a dance. At first, Car'line determines to "confront him quite calmly," but her resolve is denied her when he begins to play: "A tremor quickened itself to life in her, and her hand so shook that she could hardly set down her glass. It was not the dance nor the dancers, but the notes of that old violin which thrilled the London wife, these still having all the witchery that she had so well known of yore, and under which she had used to lose her power of independent will."[133] Here, Car'line's body is the Helmholtzian piano that trembles in response to Mop's music. The music's "thrill" leads directly to Car'line's loss of "independent will." Her body is not entirely her own; the tremor quickens "itself" to life, and her shaking hand renders her unable to even hold a glass. As the music plays, Car'line at first experiences a "paralyzed reverie" and then begins to "laugh and shed tears simultaneously."[134] The bodily effects of the music are at once unwanted and uncontrollable.

As with Maggie, Car'line's weakness in response to Mop's music renders her vulnerable not only to her initial perpetrator but also to a host of men in the hostel who beg to dance with her: "She did not want to dance; she entreated by signs to be left where she was, but she was entreating of the tune and its player rather than of the dancing man. The saltatory tendency which the fiddler and his cunning instrument had ever been able to start in her was seizing Car'line just as it had done in earlier years, possibly assisted by the gin-and-beer hot."[135] When "two or three begged her to join . . . she declined on the plea of being tired and having to walk to Stickleford, [but] when Mop began aggressively tweedling 'My Fancy-Lad,' . . . it was the strain of all seductive strains which she was least able to resist. . . . Car'line stepped despairingly into the middle of the room with the other four."[136] Car'line soon finds herself in the middle place of the cross in the reel, "the axis of the whole performance, and could not get out of it, the tune turning to the first part without giving her opportunity. And now she began to suspect that Mop did know her, and was doing this on purpose."[137] Car'line's musical convulsions lead her to be literally surrounded by male bodies in a way she cannot "get out of."

As the scene progresses, the music begins to cause her literal pain: "She convulsively danced on, wishing that Mop would cease and let her heart rest from the aching he caused, and her feet also."[138] This is literal, unwanted physical contact that causes her to ache in her "heart" and "feet." As the music continues, it "project[s] through her nerves excruciating spasms, a sort of blissful torture. . . . Car'line would have given anything to leave off; but she had, or fancied she had, no power, while Mop played such tunes."[139] Though there is "bliss" in the torture, it is still torture—an "excruciating" experience that Car'line would give "anything" to avoid. The narrator's strange intrusion—"she had, or *fancied she had*, no power"—attributes some of the fault to Car'line, suggesting that she *did*, actually, have power and that it was her mistake to think otherwise (emphasis mine). The reminders of her drunkenness throughout the scene also read all too potently as victim blaming. And yet, the narrator soon makes clear that Car'line does not, in fact, have much bodily power. As everyone else retires to their seats, Car'line continues to dance "slavishly and abjectly, subject to every wave of the melody, and probed by the gimlet-like gaze of her fascinator's open eye."[140] Suddenly, she collapses, sinking "staggering to the floor; and rolling over on her face, prone she remained. . . . Car'line was now in convulsions, weeping violently, and for a long time nothing could be done with her."[141] While Car'line is convulsing on the floor, Mop steals her daughter and takes her to America, and Car'line never sees her again. Mop weaponizes his fiddle music to deprive Car'line not only of her sexual agency but also of one of the only other agentic roles available to her (and to many Victorian women): her motherhood.

Like *The Mill on the Floss*, "Fiddler" has incited curious scholarly attempts to exonerate Aeneas and blame Car'line. Contemporary literary critics seem to be more willing to blame Car'line than Mop. Hughes writes, "Car'line's relish for music is fatally subject to a narrative treatment which insists on its ultimate recklessness."[142] Asquith suggests that while Car'line is able to remain an "honour[able]" victim because her dalliance with Mop is due to her physiological susceptibility, her own personal weakness is still at fault: "By reducing Mop's power to a simple physiological relationship, Car'line's honour is preserved: a victim of her nervous disposition rather than lax morality."[143] Yet Car'line is, first and foremost, a victim of *Mop*; however hypersensitive Car'line may be, readers must remember that Mop's music has the power to affect even a "gate-post." It is not just that Car'line lacks strength—though Hardy shrewdly shows how women are accused of just this—but that music, in the wrong hands, can be so physiologically affecting that it robs listeners of agency and subjects them to very real physical, emotional, and social harms.

While it is perhaps unsurprising to locate structurally disempowered women in Eliot's and Hardy's novels, given Eliot's proclivity for tragically fated heroines and Hardy's reputation for depicting the "catastrophes of sexuality," it is nonetheless crucial to recognize that Eliot and Hardy were capable of and invested in explicitly describing rape.[144] In the context of nineteenth-century acoustical science, the scenes in which men play music to heroines who do not want to hear it but cannot escape its material effects emerge as moments of intense violence in and of themselves, rather than vague intimations of violence or indications of more serious future harms. Moreover, in these musical rape scenes, both Eliot and Hardy draw out many of the complexities of rape and rape culture that feminist literary critics and rape theorists still discuss today: that rape can coincide uneasily with pleasure, that it can take many different forms, that it can occur in unexpected contexts (including alongside beautiful music), and that it can be so vague and ambiguous—and the "fault" of so many different people and systems—that it is often near impossible to recognize (let alone indict) the specific perpetrator(s). These texts thus prompt literary critics to envision alternative approaches to reading rape in Victorian literature. While scholars might continue to seek out metaphors and vague hints of rape, they must also look for moments in which rape is manifest on the surfaces of texts. What if readers allow that many Victorian writers were willing, indeed eager, to overtly describe sexual harm? What might such overt representations of rape—capaciously conceived—reveal about these texts' treatments of gender and sexuality? What forms of violence might readers be able to locate in their own culture if rape is no longer defined only as genital penetration? What might readers have to confront if they see rape as—like acoustic waves—diffuse and enveloping, reverberating across space and time?

Chapter Five

Orgasm in the Orchestra Box
Teleny's Musical Pornography

It is surely no coincidence that among the many code words and phrases for a homosexual man before Stonewall (and even since), "musical" (as in, "Is he 'musical' do you think?") ranked with others such as "friend of Dorothy" as safe insider euphemisms.

—Philip Brett, "Musicality, Essentialism, and the Closet" (1994)

Being a musician offered perfect plausible deniability. I could sound out gaily on the ivories while insisting that this is simply *the way you play piano*—just sounds, just vibrations, none of it admissible evidence in the court of bullies.

—William Cheng, *Just Vibrations* (2016)

For some of us, it might be that the most intense and important way we express or enact identity through the circulation of physical pleasure is in musical activity, and that our "sexual identity" might be "musician" more than it is "lesbian," "gay," or "straight." . . . If our musicalities and our sexualities are psychically next-door neighbors, how might we experience a cross-over between the two?

—Suzanne Cusick, "On a Lesbian Relationship with Music" (1994)

In the opening scene of the anonymously published pornographic novel *Teleny* (1893), the narrator, Camille Des Grieux, recollects an experience of sexual ecstasy: "That thrilling longing I had felt grew more and more intense. . . . My whole body was convulsed and writhed with mad desire. My lips were parched, I gasped for breath; my joints were stiff, my veins were swollen. . . . My brain began to reel as throughout every vein a burning lava coursed, and then, some drops even gushed out—I panted. . . . I sat there dumb, motionless, nerveless, exhausted."[1] This scene reads like many other sexual episodes, charting a familiar orgasmic trajectory of stimulation, arousal, climax, and release. Yet this moment is atypical in that the stimulus of his orgasm is neither a masturbatory act nor genital contact with another person but rather the tones of a concert piano.

Teleny begins when Des Grieux attends a charity concert at a Parisian opera house. When the pianist, René Teleny, begins to play, Des Grieux immediately responds to the charms of his music. The music ignites acute bodily sensations in Des Grieux—the convulsing, writhing, gasping, stiffening, and swelling described above. As the music "whisper[s]" in his ear, he even begins to feel a phantom hand moving across his lap: "But suddenly a heavy hand seemed to be laid upon my lap, something was bent and clasped and grasped, which made me faint with lust. The hand was moved up and down, slowly at first, then faster and faster it went in rhythm with the song."[2] In reality, though, the hand in question is placed not on Des Grieux's body but on the piano keys. It is in a concert hall, not a bedchamber, that *Teleny*'s pornographic narrative begins.

Whereas chapter 4 traced the role of acoustical science in depicting the horrors of musical rape, this chapter offers a much more reparative reading of the ties between sex and sound. Often considered one of the "first gay porn" novels, *Teleny* sets many of its sexual encounters to music.[3] This is not altogether surprising, given the close ties between music and sex in many Victorian texts (such as those discussed in chapter 4) and, especially, music's associations with "homosexuality" in the late nineteenth century.[4] While classical music has long served as a space for sexual transgression—from the castrato craze of the eighteenth century to the popularity of the opera house as a "cruising" site in the mid-twentieth century—the ties between music and sexual "inversion" were particularly potent in the Victorian period.[5] From the cello jacket that Oscar Wilde wore to the opening of London's Grosvenor Gallery, to the same-sex relationships of composers like Pyotr Ilyich Tchaikovsky and Ethel Smyth, to the musical writings of queer aesthetes like Walter Pater and John Addington Symonds, the fin de siècle witnessed a tradition of "musical homoeroticism."[6] Nineteenth-century sexologists even identified links between one's proclivity for music and their sexual preference. Evolutionary scientists like Darwin, Spencer, and Gurney linked music to sexual selection, and sexologists like Havelock Ellis, Edward Prime-Stevenson, Magnus Hirschfeld, and Edward Carpenter studied music's "peculiar[] common[ality] among inverts."[7] In the second volume of *Studies in the Psychology of Sex ("Sexual Inversion")*, Ellis cited Hirschfeld's calculation that "98 percent of male inverts are greatly attracted to music."[8] As Riddell notes, sexological studies about music and "inversion" were often used to pathologize homosexuals as possessing a "malfunction of the nerves" or "'nervous' emotionality."[9] Yet some fin de siècle thinkers, including many sexologists, also celebrated the ties between homosexuality and music, which they believed granted same-sex desire a "socially attractive heightened aesthetic capacity" and a "connect[ion] to erudite aesthetic traditions admired in Britain."[10]

Unsurprisingly, then, music and homoerotic desire frequently coincide in Victorian literature. Music most often plays a figurative role in nineteenth-century depictions of homosexuality—as a code, metaphor, allusion, or symbol for forbidden sexual practices. As musicologists Philip Brett and Elizabeth Wood argue, music's ties to the "imaginary," the "unnamed," the "unspecified," and the "unattached" have long rendered it an ideal mechanism through which to describe homosexuality.[11] Similarly, Law writes, "As an inarticulate medium with the power to stir and trouble while it seems to communicate some indefinite message, music is an ideal emblem for that which could not be named but would be recognized by those who shared in it."[12] When Wilde casts *Dorian Gray*'s murderous Alan Campbell (with whom Dorian enjoys a unique "intimacy") as a violin player, or when Katherine Bradley and Edith Cooper ("Michael Field") use the motif of birdsong to connote erotic longing, music serves as a figurative stand-in for "the love that dare not speak its name."[13]

However, while music's ineffability renders it an apt metaphor or code for illicit desire in Victorian texts in which transgressive desires must remain "unnamed" and "undescribed," its role is a bit more puzzling in a novel like *Teleny*. Whereas "respectable" Victorian novelists needed to navigate how to "describe sex and still maintain a decent distance from pornography," works of pornography, by definition, required no such euphemistic language.[14] Published anonymously and marketed to and distributed among underground networks of readers, *Teleny* hinges on frank depictions of sexual acts.[15] The authors of *Teleny* clearly had no qualms about vividly illustrating sexual encounters and the bodily sensations that accompany them, down to the "drops" of an orgasmic release.[16]

In some fin de siècle pornographic works, music provides a convenient setting for sexual encounters—a phenomenon with a clear historical basis, as music halls, balls, dances, and pleasure gardens were among the most common sites for illicit sexual encounters in Victorian society.[17] In *The Sins of the Cities of the Plain; or, Recollections of a Mary-Ann* (1881), for instance, a man dressed as "Laura" plays the piano for the narrator in exchange for sexual favors: "Now I will play you a nice piece, only I have a fancy to have you in me, and you must both fuck and frig me as I play to you."[18] In Aubrey Beardsley's *Venus and Tannhäuser* (1907), the singer Spiridion is rewarded for a successful vocal performance with an orgiastic frenzy: "The men almost pulled him to bits, and mouthed at his great quivering bottom! . . . Sup, the penetrating, burst through his silk fleshings, and thrust in bravely up to the hilt, whilst the alto's legs were feasted upon by Pudex, Cyril, Anquetin, and some others."[19] In these works, music serves as the pretext and locale for sexual ravishment.

In other examples of fin de siècle pornography, musical terminology provides rich linguistic fodder for bawdy puns and lewd double entendres. Pornographers were especially tempted by the representational possibilities of "fingering" as an act done to both musical instruments and human bodies. In *The Adventures of a Schoolboy*, written anonymously and released by the erotic book publisher William Dugdale in 1865, a man bends over a woman at a piano, which gives him "a favourable opportunity for a little fingering upon a still more delicate instrument than the one she was performing on."[20] The Austrian illustrator Franz von Bayros's drawing "Viola da Gamba" (1908), in which a cello lies next to a couple engaged in an act of manual stimulation that resembles the position of cello playing, capitalizes on the slippages between musical and genital manipulation (see fig. 5.1). Musical terminology thus facilitated vivid and playful illustrations of fleshly pleasures in several works of fin de siècle pornography.

In *Teleny*, however, music provides more than a convenient context or set of terms ripe for raunchy double entendres. Rather, music is a crucial component of the sexual encounters between Des Grieux and Teleny and a source of their erotic sensations. Some critics read music as a substitute form of contact for when Des Grieux and Teleny are unable to interact sexually. David Deutsch views Teleny's music making as a way for the men to "connect without risking public disclosure," and Riddell proposes that Teleny's piano represents "a technology for the transmission of touch . . . between queer bodies that might otherwise remain untouchable."[21] I suggest that music is a crucial source of erotic sensation in itself, not merely a conduit for the

Orgasm in the Orchestra Box | 131

Figure 5.1. "Viola da Gamba" by Franz von Bayros, c. 1908. *Source:* Varshavsky Collection, SVE-0515.2022. Used with permission.

transmission of homosexual love. Moreover, while music may offer the two men an "aesthetic experience[] that transcend[s] the flesh," as Joseph Bristow argues, music is also crucial to *Teleny*'s fleshly narrative.[22] *Teleny*—subtitled, after all, "A *Physiological* Romance"—leans on acoustical science to cast both musical and sexual encounters as intensely visceral events that ignite parallel—and equally potent—bodily sensations. Far from a mere code or meager replacement for prohibited sexual desires—Teleny and Des Grieux do have plenty of opportunities for sexual consummation throughout the novel—music appears alongside, intersects with, and is indeed at times inextricable from the novel's sexual narrative. In the orgasm scene discussed above, for instance, it is unclear whether the "writhing" and "convulsing" that Des Grieux experiences derives from the man or his music. Acousticians like Helmholtz and Tyndall would have understood that Des Grieux's reaction could just as easily have been caused by the vibratory impulses of sound waves as by the touch of a lover's hand. Sexual and sonic sensations collide to promote orgasmic pleasures; sound waves and erotic touch fuse, heightening the pleasures of both.

The question remains, then: If musical sensation is neither a code for same-sex desire nor a stand-in for forbidden sexual pleasures, but rather something that coincides and overlaps *with* sexual encounters, what is it doing in *Teleny*? If the authors are able to grant Des Grieux and Teleny plenty of bodily pleasures in the bedroom, why also focus in such depth on the sensations they experience in the concert hall? In short, why are the sex scenes so musical?

This chapter shows that the slippages between music and sex in *Teleny* are crucial to the novel's political interventions. My argument has two major strands. First, the pairings of musical and sexual response fuel the novel's defense of same-sex desire.[23] By coupling musical arousal with sexual stimulation—and rendering the two inextricable from one another—the authors of *Teleny* cast the encounters between Des Grieux and Teleny not only as gratifying but also as natural, organic, healthy, universal, and, crucially, explainable by science. This was a daring insinuation at a time when sexual "inversion" was deemed a "crime against nature" and associated with disease and degeneration.[24] As previous chapters have discussed, many nineteenth-century acoustical scientists emphasized musical response as unconscious or preconscious, automatic, and universal to most humans and animals. In the context of acoustical science, then, Des Grieux's parched lips, quick breath, and stiff joints in the scene above are as much organic and involuntary responses to the music he hears as they are signs of his sexual arousal. In *Teleny*'s musical scenes, sexual pleasure transforms from a licentious indulgence in a dark street corner into a sensation akin to those induced by acoustical demonstrations at the Royal Institution. Moreover, the novel casts Des Grieux's bodily responses to music as universal sensations that everyone in the concert hall experiences. In doing so, the authors of *Teleny* push back against nineteenth-century sexological discourses that framed homosexuals' musical proclivities as components of their disordered nervous systems.[25] Des Grieux's musical-sexual sensations are natural and even *normal*—shared by all listeners.

Second, I argue that alongside *Teleny*'s crucial defense of same-sex desire is a much broader imagination of queer erotic possibilities. What if readers take seriously the possibility that Des Grieux's opening orgasm might really be in response to the music itself? *Teleny* can be read as a narrative of desire between men and also as an exploration of music *itself* as a sexual partner, one that is shared among a multitude of players and listeners. In line with queer theoretical work that conceptualizes sex as an act that can occur outside of reproductive or even genital realms—including Taylor's framework of "aural sex"—*Teleny*'s sexual narrative is even more capacious than critics have allowed.[26] This is not to minimize the novel's representation of same-sex desire but rather to suggest that its authors explored an even wider range of sexual possibilities than critics often acknowledge, including those shared between lovers, among players and listeners, among bodies in a concert hall, and among humans and vibrations of music in the air.

"Natural Tastes": *Teleny*'s Defense of Same-Sex Desire

As Ed Cohen and Joseph Bristow argue, *Teleny* foregrounds same-sex desire as natural and healthy.[27] The interlocutor refers to Des Grieux's "natural tastes," and Des Grieux repeatedly insists on what Benjamin Kahan would call the "congenitality" of his sexual preferences.[28] Des Grieux describes himself as "predisposed to love men and not women," relates that his "hankering for males" began in "childhood," and insists that he was "born a sodomite."[29]

The novel emphasizes the organicity of same-sex desire not only through the language of congenitality but also, perhaps more unexpectedly, through the language of acoustics. The authors repeatedly invoke physical and physiological acoustics to describe the men's bodily pleasures as they listen, play, and have sex. In doing so, they attribute Des Grieux's and Teleny's bodily sensations—both musical and sexual—to the automatic and universal actions of the nervous, musculoskeletal, and circulatory systems. By attaching sexual sensations to musical ones, the authors of *Teleny* show that same-sex desire was just as natural and defensible as was musical appreciation.

Teleny's investments in physiology and appeals to "natural" bodily phenomena are certainly disorienting in light of the violent histories of Victorian biological science, as well as more recent feminist and queer critiques of essentialism by thinkers who put pressure on "born this way" narratives of sexual identity that make little space for contingency and variability and that can be trans-exclusionary.[30] Victorian scientists and pseudoscientists, after all, most often weaponized ideas of "naturalness" against those marginalized by sex, sexuality, race, class, gender, and ability, whom scientists, doctors, and other intellectuals often deemed naturally unintelligent, naturally primed for hard labor, or dangerously *un*natural.[31] As Dustin Friedman writes, Victorian scientists were especially "intent on reducing queers to mere bodies, passively in thrall to diseased impulses beyond their control."[32] The same logic that *Teleny* uses to describe bodies as innately musical was also, especially in the nineteenth century, used to describe certain bodies as innately undeserving of basic human rights. Indeed, *Teleny* at times recapitulates violently essentialist rhetoric to convey "medicalised" discourses about "degeneration," to rehearse Orientalist thinking, or to cast women as grotesquely *hyper*natural—slimy, scabby, oozing, and cadaverous—and subject to consumption, rape, and suicide.[33]

And yet, the writers of *Teleny* found enormous potential in their "queer use" of acoustical science as a mechanism to defend, not demonize, homoerotic desire.[34] In *Teleny*, same-sex encounters align not with syphilitic illnesses or mental neuroses but with beating hearts, moving limbs, sweating skin, and vivid memories. In this way, *Teleny* resonates with recent work by feminist and queer theorists who recognize biological science as an unexpected site for queer politics.[35] For *Teleny*'s authors, the notion that bodies acted, operated, and responded according to scientific laws proved to be a valuable tool, particularly for those whom science was often mobilized *against*. On the heels of the Labouchère Amendment and the eve of the Wilde trials, which both further criminalized homosexuality (and did so especially publicly and violently), there was perhaps nothing so powerful as to claim, as the authors of *Teleny* did, that same-sex desire, like musical enjoyment, was one of the most natural things in the world. Loving another person of the same sex was as organic and life-sustaining as hearing a symphony or remembering a long-forgotten tune.

A musical encounter propels the sexual relationship between Des Grieux and Teleny. During their first meeting, Teleny tells Des Grieux that he is a "sympathetic listener"—"a person with whom a current seems to establish itself; someone who feels, while listening, exactly as I do whilst I am playing, who sees perhaps the same visions as I do."[36] Critics mostly focus on the electricity metaphor in this passage or on the references to the lovers' shared "visions" as a nod to contemporary trends in psychical research or telepathy.[37] However, Teleny's comments are equally indebted to nineteenth-century acoustical science, particularly Helmholtz's theory of sympathetic vibration. While Teleny's statement about the "current" between them may seem merely to metaphorize their electric kinship, in the context of Victorian acoustical theories that uncovered the ability of human bodies to vibrate "sympathetic[ally]" in response to sound, this moment testifies to an actual instance of physical contact between the two men. In acoustical terms, Teleny and Des Grieux are quite literally vibrating at the same frequency. The attraction of male lovers is just as explainable as the physics of sound.

Des Grieux's most "sympathetic" listening experience occurs, of course, in the concert hall orgasm scene discussed at the beginning of this chapter:

> The pianist's notes just then seemed murmuring in my ear with the panting of an eager lust, the sound of thrilling kisses. . . . The music just then seemed to whisper. . . . That thrilling longing I had felt grew more and more intense . . . my whole body was convulsed and writhed with mad desire. My lips were parched, I gasped for breath, my joints were stiff, my veins were swollen. . . . My brain began to reel as throughout every vein a burning lava coursed, and then, some drops even gushed out—I panted—I sat there dumb, motionless, nerveless, exhausted.[38]

Critics have read this as an instance of metaphysical connection between the men (based on the "visions" of Egypt and Spain that the men conjure during the performance), as a "spatially displaced" encounter between otherwise "untouchable" bodies, or as a depiction of masturbation in which the hand on the lap is in fact Des Grieux's own.[39] Yet the passage also reveals Des Grieux's concrete experience of physiological arousal by both Teleny and his music. In the context of acoustical scientific discourses about music's ability to quiver the nerves inside the human ear, this moment takes on especially vivid physiological import. The "murmuring" in Des Grieux's ear serves not as a metaphor but rather as an indication of the actual ways in which Teleny's music titillates Des Grieux's auricular anatomy. Moreover, the descriptions of Des Grieux's tense muscles, stiff joints, and "swollen veins" echo key principles of physiological acoustics, namely that the sounding of tones in the ear can cause kinesthetic arousal throughout the body.[40] Teleny's concert hall orgasm looks a lot like the expected response of the human body to music.

The overlap of musical and sexual response continues as Des Grieux describes the kinds of music that Teleny plays. As he reflects to his unnamed interlocutor, the Hungarian music that Teleny performs is uniquely "sensuous" because of its minor scales, "rare rhythmical effects," and rules of harmony that make it "ja[r] upon our ears."[41] Des Grieux explains,

> You cannot disconnect him from the music of his country; nay, to understand him you must begin by feeling the latent spell which pervades every song of the Tsigane. A nervous organisation—having once been impressed by the charm of a tsardas—ever thrills in response to those magic numbers. Those

strains usually begin with a soft and low andante, something like the plaintive wail of forlorn hope, then the ever changing rhythm—increasing in swiftness—becomes "wild as the accents of lovers' farewell," and without losing any of its sweetness, but always acquiring new vigour and solemnity, the prestissimo—syncopated by sighs—reaches a paroxysm of mysterious passion, now melting into a mournful dirge, then bursting out into the brazen blast of a fiery and warlike anthem.[42]

Critics rightly focus on this moment as a reflection of Victorian exoticization (and even xenophobic suspicion) of "Eastern" or "Asiatic" music.[43] But this moment also further muddles the distinctions between musical and sexual response. The passage goes to great lengths to merge the performer and the music; indeed, Teleny and his music "cannot [be] disconnect[ed]." The litany of musical terms ("accents," "rhythm," "andante," "syncopated," "dirge," "anthem") specifically links Des Grieux's "writhing" and "convulsing" to the music itself as much as to the performer. Des Grieux, a discerning listener with in-depth knowledge of the history and cultural conventions of classical music—so much that the interlocutor complains about Des Grieux's overuse of "technical terms"—attributes his quivering body to the specific rhythmic and dynamic intricacies of a tsardas, a Hungarian folk dance that has an especial ability to "impress" a "nervous organisation." Language like "ever thrills" and "usually" suggests that such responses *frequently* coincide with Hungarian music and are its expected effects. Moreover, while the neurological language in this passage—"nervous organisation" and "paroxysm"—might evoke contemporary associations between sexual "inversion" and nervous disease, the scene can also be read as a nod to acoustical theories about the role of the nervous system in musical sensation.[44] Des Grieux's nervous system is specifically "organised" to respond to Teleny and his music.

The novel further emphasizes the naturalness of Des Grieux's sexual desires in the moments in which he links his sexual passion for Teleny to the neurological process of getting a song stuck in one's head, or what is now often called an "earworm":[45]

> The more I tried not to think of him, the more I did think. Have you in fact ever heard some snatches of a half-remembered tune ringing in your ears? Go where you will, listen to whatever you like, that tune is ever tantalising you. You can no more recollect the whole of it than you can get rid of it. If you go to bed it

keeps you from falling asleep; you slumber and you hear it in your dreams; you wake, and it is the very first thing you hear. So it was with Teleny; he actually haunted me, his voice—so sweet and low—was ever repeating in those unknown accents. Oh! friend, my heart doth yearn for thee.[46]

This moment might at first seem to cast Des Grieux's attraction as a dangerous or unhealthy obsession. And yet, the earworm analogy frames Des Grieux's desire more as an innocuous, even pleasurable—and completely normal and common—neurological phenomenon. Nineteenth-century scientists described earworms as inevitable and universal physiological processes. Gurney, for instance, wrote extensively about musical memory and described music's power of "getting into the blood and clinging to the memory."[47] For him, "a person may be haunted by music in the midst of and without interruption to the busiest and most opposite avocations."[48] In this context, then, Des Grieux's desire for Teleny emerges not as a tic of a diseased brain but as a common neurological occurrence. "Tantalizing" and "haunted" suggest that Des Grieux is ambivalent about this earworm. He is partially annoyed that the tune disrupts his sleep and that he can only recollect "snatches" of the tune. And yet, the earworm also provokes extreme physical pleasure, enabling Des Grieux to recollect Teleny's "sweet and low" voice. The term "tantalizing" connotes torment but also teasing, excitement, taunting, and *almost* untenable desire. Although earworms were (and still are) often framed as annoyances or even forms of torture—the composer Robert Schumann allegedly went mad after being unable to rid his brain of the sounding of a specific series of tones—for Des Grieux, the sonic memory of Teleny's music is a "sweet" haunting.[49] Moreover, the authors of *Teleny* foreground the earworm as inevitable and natural; though Des Grieux tries not to think of Teleny, the tune keeps ringing in his ears, as if willing him to recall his memory of his lover's music and relive the pleasures he once experienced. He cannot help but think of anything else.

As the novel progresses, the earworm becomes an even more important source of sexual pleasure and memory preservation for Des Grieux. In a later scene, he hears the tones of Teleny's voice replaying in his head: "[Teleny] then began to whisper words of love in a low, sweet, hushed and cadenced tone that seemed like a distant echo of sounds heard in a half-remembered ecstatic dream. . . . I can even now hear them ringing in my ear. Nay, as I remember them again, I feel a shiver of sensuality creep all over my body, and that insatiable desire he always excited in me kindles my blood."[50]

This passage is certainly not without ambivalent words (such as "shiver" and "creep"), and Des Grieux's desire is frustratingly "insatiable." However, the use of the present tense—words like "even now," "feel," "kindles," and "remember"—demonstrates that these physical experiences are still occurring in Des Grieux's body. Like the gramophones, phonographs, and other sound technologies emerging at the end of the nineteenth century, earworms enabled the human body to act as a mechanism for sound preservation and reproduction. Through musical memory, Des Grieux and Teleny's erotic encounters can, to some extent, persist. Des Grieux's erotic life is not fully relegated to the past; music offers him a way to relive his sexual experiences for later sustenance. At a time when Victorian moralists denigrated same-sex relationships for their *in*ability to reproduce, the auditory actions of Des Grieux's own brain and body allow him to re-create his erotic encounters. Same-sex desire was not degenerate but life-sustaining and even reproducible through sonic memory.

The narrative also describes at length Des Grieux and Teleny's genital encounters and their budding romance. Crucially, the authors use similar physiological language to describe the men's sexual and musical relationships. In one scene, for instance, Des Grieux and Teleny entwine their fingers: "All the blood vessels of my member were still strongly extended and the nerves stiff, the spermatic ducts full to overflowing."[51] This passage harkens back to the musical arousal Des Grieux feels "throughout every vein" upon first hearing Teleny's rhapsodic melody. He even describes several of his sexual encounters with Teleny as instances of sympathetic vibration. In one moment, he relates how "that touch . . . vibrated through all my body, giving all the nerves around the reins a not unpleasant twinge."[52] Elsewhere, he notes that "all my nerves were vibrating as if set in motion by some strong electric current."[53] By mapping the language of vibrating nerves, circulatory arousal, and tingling skin—sensations elsewhere directly linked to musical response—onto descriptions of sexual encounters, the authors invite readers to consider that Des Grieux's sexual sensations might be just as scientifically explainable as his musical ones; both are predictable actions of the central nervous system.

Thus far, one could argue that the merging of neurological responses with music and sex merely works to recapitulate homophobic Victorian discourses about the diseased "nerves" of musicians/"inverts." Just because Des Grieux's musical response is "natural," after all, does not mean that it would be considered "healthy" or "normal" among Victorian sexologists. However, the novel goes to great lengths to cast Des Grieux's musical-sexual responses as *universal* sensations experienced by everyone in the concert hall. As many nineteenth-century acousticians argued, physiological responses to music were common among all humans (though for some, as previously discussed, this "universality" was severely limited by race and class[54]). Havelock Ellis believed that even the "unmusical subject responds physiologically, with much precision, to musical intervals," and Chomet wrote, "In all [mammals and humans], without exception, [music] will arouse sensations and impulsive emotions of some kind."[55] By emphasizing that many listeners share Des Grieux's responses to Teleny's music, the authors leave open the possibility that his *sexual* sensations—so inextricable from the musical ones throughout the text—may also be more common than readers might think.

It is not just Des Grieux who is aroused by Teleny's music. At the beginning of the opening concert hall scene, Des Grieux tells the interlocutor,

> He sat down and began to play. I looked at the programme; it was a wild Hungarian rhapsody by an unknown composer with a crackjaw name; its effect, however, was perfectly entrancing. In fact, in no music is the sensuous element so powerful as in that of the Tsiganes. . . . If you have ever heard a tsardas, you must have felt that, although the Hungarian music is replete with rare rhythmical effects, still, as it quite differs from our set rules of harmony, it jars upon our ears. These melodies begin by shocking us, then by degrees subdue, until at last they enthral us.[56]

Des Grieux attaches his sensations to those of his fellow concertgoers. The collective pronouns ("jars upon *our* ears" and "enthral *us*") are illustrative. Des Grieux is not alone in his physiological response; rather, it is expected that most listeners (at least most English ones) will be "enthral[led]" by Teleny's music. Des Grieux further emphasizes the commonality of his experience by appealing to the interlocutor's own experiences ("if you have ever heard," "you must have felt that"). The phrase "must have," moreover, further emphasizes this kind of response as inevitable, not unnatural.

Even as his orgasm builds, Des Grieux connects his experience to that of the collective: "My lips were parched, I gasped for breath; my joints were stiff, my veins were swollen, yet I sat still, like all the crowd around me."[57] The passage here can certainly be read as suggesting that that the crowd was, in fact, still, and that Des Grieux had to make every effort to appear like them. Yet, particularly with the litany of commas that muddle the sentence's referents, the passage also leaves open the possibility that "all the crowd" might have had stiff joints and swollen veins. The collective eroticism of this scene is heightened in the 1934 French translation of the text by Charles Hirsch, who claimed that the authors of *Teleny* used his London bookshop, the Librarie Parisienne, as an exchange site for the working manuscript.[58] In Hirsch's edition, the phrase reads, "comme ceux qui m'entouraient" ("like those around me" or "like all those who surrounded me").[59] The use of the verb *entourer* ("to surround") more vividly situates Des Grieux as one of many bodies in the concert hall. His heart and limbs are just some of the many that are palpitating and quivering throughout the concert hall, especially given the "thundering applause of the whole theatre" that follows the performance.[60] In both editions, then, the authors of *Teleny* leave readers with a tantalizing possibility: that Des Grieux was not the only one vibrating in his seat, and perhaps not even the only one releasing a few gushing "drops."

The emphasis on the universality of Des Grieux's sensations persists throughout the novel. At a later concert, Des Grieux reveals, "I was thoroughly overcome, whilst the whole crowd was thrilled by the sweet sadness of his song."[61] Elsewhere, when describing Teleny's rising fame, Des Grieux tells the interlocutor that "all those whose blood was not frozen with envy and age were entranced by that music. His name, therefore, began to attract large audiences, and although musical critics were divided in their opinions, the papers always had long articles about him."[62] Teleny's performance arouses public fascination, stimulating the "blood" of all of his listeners, and the "entranc[ing]" qualities of his playing are sanctioned in the public sphere of print culture.

Even Teleny's earworms are described as universal. Note the persistence of the second-person "you" in the initial earworm passage: "Have you in fact ever heard some snatches of a half-remembered tune ringing in your ears? Go where you will, listen to whatever you like, that tune is ever tantalising you. You can no more recollect the whole of it than you can get rid of it. If you go to bed it keeps you from falling asleep; you slumber and you hear it in your dreams; you wake, and it is the very first thing you hear." Des Grieux's appeal to the "you"—at first directed at the interlocutor but then transformed into a more collective moniker ("you can no more recollect")—casts the earworm not as a form of madness or disease but as a common phenomenon. Des Grieux's earworm—and thus his sexual passion—is something to which everyone can relate. Such moments call readers to ponder whether Des Grieux's response is as "abnormal" or "unnatural" as Victorian sexologists would have thought. If his musical arousal—so tied to his sexual arousal—is "degenerate," then it is a degeneration shared by all in the concert hall.

Aural Sex: *Teleny*'s Queer Erotics

At the Feminist Theory and Music Conference in 1991, Cusick issued her now-famous provocation:

> What if music IS sex? If sex is free of the association with reproduction enforced by the so-called phallic economy (and it is, remember, exactly so for people called homosexual, as it has become in the last thirty years for people called heterosexual who practice contraception), if it is then *only* (only!) a means of negotiating power and intimacy through the circulation of pleasure, what's to prevent music from *being* sex, and thus an ancient, half-sanctioned form of escape from the constraints of the phallic economy. . . . If music IS sex, what on earth is going on in a concert hall during, say, a piano recital?[63]

Cusick's comments have sparked ongoing musicological discussions of the "queer possibility" that humans might actually have sex *with* music, particularly if sexuality is understood as "a way of expressing and/or enacting relationships of intimacy through physical pleasure shared, accepted, or given."[64] This line of thinking, for instance, underscores Taylor's claims that "the ear can be an erogenous zone" and that music "may also be a powerful sexual stimulant that brings [people] to a point of sexual gratification."[65]

Though a century removed from *Teleny*, these musicological discussions prompt readers to examine Des Grieux's narrative from a different angle. What *is* going on at Teleny's piano recital? In this closing section, I submit that the ties between musical and sexual sensation in *Teleny* extend even beyond the novel's defense of same-sex love. What if readers flirt with the possibility that Des Grieux—and his fellow audience members—are just as sexually aroused by the music itself as by Teleny the man? If sexual responses can be musical, can musical responses be sexual? What if *Teleny* is even queerer than critics give it credit for?

The proposition that "music IS sex" allows readers to interpret *Teleny* as imagining a startlingly wide range of sexual practices that extend far beyond same-sex desire. Many of the novel's musical-sexual scenes can be read as depictions of "aural sex"—moments that are just as invested in the relationships between humans and music as between humans themselves.[66] In several of the passages quoted above, it is the *music*, not the *musician*, that is the subject of the sentences and the catalyst of the audience members' physiological sensations. In the concert hall orgasm scene, for instance, it is not just the *pianist* but the "pianist's *notes*" that are described as spurring Des Grieux's response: "The pianist's notes just then seemed murmuring in my ear with the panting of an eager lust, the sound of thrilling kisses . . . the music just then seemed to whisper."[67] The music and the music*ian* merge grammatically as the notes are attributed to the pianist as well as the piano music. The overlap between music and musician persists as the passage continues. Des Grieux tells the interlocutor, "He in beauty, as well as in character, was the very personification of this entrancing music. As I listened to his playing I was spellbound; yet I could hardly tell whether it was with the composition, the execution, or the player himself."[68] While writers more commonly use musical language to metaphorize human experiences, here it is the *man* who "personif[ies]" the *music*. Des Grieux cannot precisely name the source of his pleasure—whether the piece, the performance, or the performer. Erotic sensation not only coincides with but is inextricable from musical response.

The intimations of "aural sex" are even more extreme in the French edition. The orgasm passage reads, "Les notes du pianiste murmuraient à mes oreilles, avec le halètement d'une fiévreuse concupiscence, le bruit d'une roulade de baisers."[69] The phrase "roulade de baisers" directly merges music and sex. A "roulade" is a musical "run" or "trill," and a "baiser" is a "kiss," so in translation the phrase reads "a run of kisses" or a "trill of kisses." Moreover, since the seventeenth century, "baiser" has been a common slang term for an act of copulation (often translated as "fuck" or "screw").[70] Sex and music are thus impossible to disentangle here; sex can be musical (kisses or fucks that trill), and music can be sexual (trills that act like kisses or fucks). In the world of *Teleny*, music itself has the capacity to whisper, pant, kiss, and fuck. When Des Grieux later tells his mother that "it must have been [the] concert that upset my nerves," he might not be hiding his sexuality but rather describing his phenomenological experience of music listening.[71]

Within an "aural sex" framework, the emphasis on the universality of Des Grieux's musical responses becomes even more provocative. The fact that he acts "comme ceux qui m'entouraient" hints that his experience of musical erotics is shared by all of the listeners in the concert hall. From this angle, the concert hall of Des Grieux's memory begins to look a lot less like a space of Victorian comportment and a lot more like a realm of communal, orgiastic pleasure.

In this way, *Teleny* anticipates queer-theoretical conversations that explore forms of sex that extend beyond reproductive or even genital realms (as Cusick does above, with her definition of sex as "relationships of intimacy through physical pleasure shared, accepted, or given"[72]). In their rereadings of Foucault, for instance, queer theorists Tim Dean and David Halperin emphasize that not all pleasure is sexual and not all sex is genital. Halperin argues that Foucault viewed sadomasochistic (S/M) practices as important not because they "detach[] pleasure from all acts of a conceivably sexual nature" but because they "detach[] sexual pleasure from genital specificity, from localization in or dependence on the genitals. . . . [S/M] involves the eroticization of nongenital regions of the body, such as the nipples, the anus, the skin, and the entire surface of the body."[73] Halperin's reinterpretation of Foucault prompts readers to think about where *else* on the body sexual pleasure can be experienced—a query to which Cusick, Taylor, and the authors of *Teleny* would likely respond: the ear. While Halperin emphasizes the variety of erogenous body parts, Dean underscores the range of experiences that can be defined as sexual, such as how "some people 'love literature' *in exactly the same way* as others love sex."[74] In *Teleny*, loving music and loving sex might very well be the same thing.

As important as it is to locate *Teleny* within a specific history of homosexuality, then, critics should also acknowledge its broader, and perhaps bolder, intimations about the wide range of erotic possibilities available even in a society so committed to restricting sex. In the world of *Teleny*, there are many ways to have sex: some genital, some aural; some in the bedroom, some in the concert hall; some with pulsing human bodies, some with vibrating sound waves.

Part Four

Intimacies

Chapter Six

Fiddle Feelings

Human-Instrument Intimacies in Dickens, Eliot, Trollope, and Hardy

> Zoe always kept the heat turned up in her apartment so that she could wear as little clothing as possible. She liked feeling the vibrations of her instruments, she said. She liked feeling the vibrations of the earth underneath her and the air around her. . . . Zoe joked—or maybe it wasn't a joke—that her first sexual experience had been with her cello.
>
> —Gabrielle Zevin, *Tomorrow and Tomorrow and Tomorrow* (2022)

In a rather minor moment in Charles Dickens's *Dombey and Son* (1848), readers meet the "elderly bachelor" Mr. Morfin and his main "companion": his cello.[1] For Morfin, the cello provides an alternative to the cruel world of human relations and capitalist speculation that dominates his professional life as an "officer of inferior state" at the House of Dombey.[2] He often returns home from work to "calm his mind by producing the most dismal and forlorn sounds out of his violoncello before going to bed."[3] In one scene, Morfin points to the cello on his sofa and tells Harriet Carker, "I have been here, all day. Here's my witness. I have been confiding all my cares to it."[4] Morfin "scrap[es] consolation out of its deepest notes" and "play[s] over and over again, until his ruddy and serene face gleamed like true metal on the anvil of a veritable blacksmith."[5] The cello, "big with the latent harmony of a whole foundry full of harmonious blacksmiths," responds "in unison with his own frame of mind, glid[ing] melodiously into the Harmonious Blacksmith."[6]

As the previous chapters have established, musical instruments resound throughout Victorian literature. Yet the instruments themselves are rarely the foci of literary texts or the criticism surrounding them. Scholars most often read instruments as metonyms or symbols for characters' personalities, class positions, or gender identities, as when a parlor piano indexes a household's middle-class status, or when the possession of a violin signals a woman's gender deviance. Sometimes instruments appear as conduits for human relationships—objects *through which* players affect or communicate with listeners. For instance, Jeremy Chow argues that the lute in Ann Radcliffe's *The Romance of the Forest* (1791) serves as a tool that "makes female intimacy possible."[7] Similarly, Riddell claims that fin de siècle queer texts depict musical instruments as "conduits for the often unspoken transmission of desire between displaced bodies" that are forbidden to interact in real life.[8]

Instruments do sometimes occupy more central places in literary works, most often as beloved objects or recipients of their players' deep affections. As discussed in chapters 2 and 3, Francis's heroine Valérie refuses to let her fiddle case out of her sight, while Thomas's Laurence refers to her violin as her "loyal old comrade."[9] In Herman Melville's *Pierre* (1852), a character named Isabel makes a "loving friend" of her guitar, while in the anonymously published short story "The Maker of Violins" (1900), the Chinese instrument maker Shing Poon "loves his violins as most fathers love their children."[10] The violinist Marion Scott wrote an entire collection of poems titled *Violin Verses* (1905), which includes a series of sonnets addressed to violins and is dedicated to Scott's own "Betts Strad" and "all [her] other friends among fiddles."[11]

But what if musical instruments are not merely conduits for human relations, nor passive recipients of musicians' one-way affections, but rather erotic partners in their own right? What if instrument playing does not always reflect a human's yearning for contact with another *human* but rather serves as a corporeally satisfying experience in and of itself? The passage from Gabrielle Zevin's 2022 novel *Tomorrow and Tomorrow and Tomorrow*, quoted in this chapter's epigraph, clearly indexes this possibility—but so too do many canonical nineteenth-century novels. As the scene from *Dombey and Son* demonstrates, Morfin's cello is not merely a symbolic object nor a passive vessel of his affection; it is instead his confidant and interlocutor and even a stand-in for a child (the object of his "paternal affection") and a surrogate partner (the "companion of his bachelorhood").[12] The cello is a vital companion that possesses its own vigorous body and has the capacity to resound "in unison" with Morfin's own thoughts. The act of cello playing grants Morfin both emotional consolation and physical release. He handles the instrument "with great tenderness and care," and the act of playing invigorates his entire body, making his face "gleam," ostensibly with the sweat he produces as he "scrap[es] away" with his bow.[13] The most fulfilling aspect of Morfin's earthly existence—his musical life—hinges on his physical ties to his instrument.

Eliot's *The Mill on the Floss*, Anthony Trollope's *The Warden* (1855), and Hardy's "Haunting Fingers" (1921) join *Dombey and Son* in depicting human-instrument encounters that grant both the characters and their instruments vibrant erotic experiences and sustaining intimate connections. As I discuss in further detail below, it was no accident that these authors were so preoccupied by the corporeal ties between humans and instruments, given that Victorian acoustical scientists and music practitioners increasingly recognized musical performance as a fundamentally physiological process. Not only did music travel in waves and act on bodies and things, but music *making* necessitated players' extensive bodily output and sustained physical contact with their instruments, which needed to be held and handled in certain ways and required particular muscular actions in order to sound. Moreover, acoustical scientists often framed musical instruments as possessing their own kinds of agency. Instruments could alter the bodies of musicians, who spent hours training their fingers and shaping their mouths to fit keys and mouthpieces. Acousticians also described musical instruments as possessing their own "muscles" and "nerves," capacities for sentience, and potentials for dynamic action—including the capacity to echo beyond their players' control.

Though the texts treated here do not deal as explicitly with issues of gender and sexuality as those discussed in previous chapters, they nonetheless foreground feminist and queer politics. The human-instrument encounters described by Dickens, Eliot, Trollope, and Hardy make space for the subjectivities of material objects, imagine erotic experiences between humans and nonhumans, and depict forms of kinship and intimacy that far surpass traditional social arrangements.

First, and perhaps most obviously, these texts invest musical instruments with startling capacities for agency and sentience.[14] While several of the works discussed in previous chapters flirt with nonhuman possibilities, especially when characters experience intimacy with music itself (when Valérie twiddles her fingers in the concert hall, or when Des Grieux feels just as aroused by Teleny's music as by his body), the texts considered here move even farther away from anthropocentric realms as they depict musical instruments as "vibrant matter" and make concrete the erotic potential of musical performance even apart from romantic or sexual relations.[15] New-materialist and posthumanist work in Victorian studies often focuses on "metonymic search[es]" into the contexts and histories of objects referred to in novels and their situated meanings in Victorian culture, but the musical instruments that Dickens, Eliot, Trollope, and Hardy depict are more than just metonyms.[16] Drawing on acoustical science, which highlighted the vibratory and resonant capacities of musical instruments, these authors invest instruments with what theorist Jane Bennett calls "thing-power," or "the strange ability of ordinary, man-made items to exceed their status as objects and to manifest traces of independence."[17] In the texts discussed here, pianos, cellos, violins, and horns possess their own bodies and experience their own sensations, affects, and responses to external stimuli. For Dickens, Eliot, Trollope, and Hardy, musical instruments are not just lifeless physical objects but resonant bodies that vibrate under their players' hands and lips.

Several new-materialist and posthumanist theorists argue that it is a fundamentally feminist and queer project to conceptualize the agency of nonhuman things. What Rosi Braidotti calls "posthuman feminism" "relinquishe[s] the liberal vision of the autonomous individual"—a vision especially beloved in the Victorian period—by "produc[ing] an expansion of the notion of corporeal matter to include non-anthropomorphic and non-human bodies" and "recompos[ing] . . . the human along posthuman axes of multi-scalar relational interconnections to non-human others."[18] Similarly, Patricia MacCormack writes that both queer and posthumanist theory "interrogate the arbitrary nature of systems of power masquerading as truth. . . . Queer theory and the posthuman mobilise and radicalise the here and now through desire, pleasure and pure potentiality."[19] Though the works of Dickens, Eliot, Trollope, and Hardy are rarely read in either queer or posthuman contexts, they nonetheless achieve the kinds of radical "recompositions" that Braidotti discusses, reorienting scholars' focus away from the singular, liberal human subject cherished in Victorian culture and toward the possibility that other beings might be vibrating just outside of view.

Not only do Dickens, Eliot, Trollope, and Hardy foreground the vibrancy of musical objects, but they also underscore the erotic sensations that music making offers both players and instruments. In their depictions of human-instrument encounters, these four authors locate an expanded field of erotic possibilities based not on sex, romance, or reproduction but on intimate bodily contact between humans and objects. For the players and instruments discussed here, the acts of playing and being played *upon* are not mere replacements or "conduit[s] for . . . transmission" for forms of sexual pleasure or social relations otherwise inaccessible to them; they are rather the most meaningful erotic activities both groups ever experience.[20] The instruments and humans feel each other's touch and breath; sense each other's gestures of pressure and release; and exchange bodily substances (skin cells, spit, oils, sweat, rosin, glue, finish). They are malleable, coextensive beings that shape and are shaped by each other.

While the human-instrument interactions discussed here do not intersect with sex in the same ways as the musical encounters addressed in chapters 4 and 5 do, they are nonetheless deeply erotic in the capacious sense of the term introduced by Lorde and further theorized by feminist, queer, and asexuality scholars to describe experiences that "supersede[] reproduction but [are] still anchored in the body."[21] Indeed, the fact that these experiences occur beyond the human-centric world might even make them all the more erotic. As Eileen Joy argues, there is a particular "pleasur[e] and enjoyment" in a "heightened contact with the world itself, in all of its extrahuman (yet still co-implicate) vibrations."[22] Similarly, Katherine Behar writes that "fomenting erotic fusion with an object . . . is a creative, generative act."[23] As several musicologists and performance theorists note, musical performance provides a particular kind of "extrahuman" pleasure. Cusick describes the uniquely erotic gratification of using her "brain and hands and feet" to play the organ, while Pedro Rebelo writes that "the performer/instrument relationship can be seen as an erotic one" because the haptic contact between player and instrument involves a "vertiginous . . . dissolv[ing of] object and subject."[24] The literary texts discussed here capture the "vertiginous" pleasures of the enmeshments of human and nonhuman worlds through music.

Finally, in depicting music making as the pinnacle of erotic experience for both players and instruments, Dickens, Eliot, Trollope, and Hardy imagine what feminist and queer theorists refer to as "queer kinship" relations, "affiliative" networks, or "queer orientations," which, as Ahmed writes, "put within reach bodies that have been made unreachable by the lines of conventional genealogy."[25] Depictions of instrument playing provide unexpected sites for Victorian writers to envision entirely new kinds of intimate social formations, ones that departed from the ideals of heterosexual marriage and the nuclear family that were so naturalized in Victorian culture—and, crucially, most often central to the kinds of realist works written by canonical authors like Dickens, Eliot, Trollope, and Hardy. Vibrant—and vibrational—encounters between humans and instruments offer glimpses of alternative social relations that are just as (if not more) gratifying as those found in the strictly human world. The relations between Maggie and her piano, Septimus and his cello, and Hardy's instruments and their players represent strikingly capacious forms of intimacy and propel readers to imagine the world as animated with humans *and* things that sense, feel, and sound. Thus, while Dickens, Eliot, Trollope, and Hardy do not depict cross-dressing violinists or pornographic concert hall encounters, they nonetheless offer equally radical reimaginations of ways of being in the world.

"Corporeal Co-Dependence": Human-Instrument Relations in Victorian Acoustical Science

In an article about the twentieth-century pianist Glenn Gould, musicologist Paul Sanden describes the extramusical sounds he hears in Gould's recordings: acoustic traces of Gould's fingers hitting the keys, "the *noise* of the piano as it is played."[26] Sanden writes, "I find these noises significant and indicative of the vital corporeality of piano performance. . . . Such recognition of the physical relationship established between Gould's body and that of his piano . . . emphasizes a sort of corporeal co-dependence between musician and instrument. In other words, without one another, both of these bodies remain mute. Music sounds only when they come together in performance."[27] In recent years, musicologists have increasingly attended to the "corporeal co-dependence" between players and their instruments. Scholars in the fields of feminist and queer musicology in particular have emphasized the need to veer away from conceptions of music making that center on the musician's "command and control" of the instrument and instead focus on musical performance as a "special, haptic relationship" or "physical coupling" between player and instrument.[28] To some contemporary musicologists and music scientists, music making represents a kind of "cybernetic human-machine extended system," a "feedback loop system," or an "exchange of mechanical energy."[29] Instruments not only respond to human breath and touch but also possess bodies of their own. Piano keys dampen and release; violin strings vibrate in their players' hands; brass mouthpieces buzz and tickle their players' lips; woodwind instruments expand and contract due to environmental pressure. The acts of practicing and playing an instrument can also fundamentally reshape the player's body. As the musicologist Maria Delgado recently wrote of the pianist Carles Santos, "[He] often configures the instrument [the piano] as a prosthetic limb, a configuration indelibly bound up with his own movements: recognition of the implications of the two or three hours he spends in daily practice."[30] Similarly, Rebelo describes the "feedback loop" involved in playing a wind instrument: "The player constantly adapts and configures throat position, vocal cavity and breath depending on the resistance of the tube, mouthpiece or reed."[31] To invoke Ahmed, music making requires players and instruments to become "oriented" to each other—to adapt to, affect, and even alter one another's bodies.[32]

Though this focus on the embodied player-instrument relationship marks a relatively recent trend in twenty-first-century musicology, many Victorian thinkers were similarly aware of the "physical couplings" involved in music making. While the "virtuoso craze" of the 1830s and 1840s treated performers with mystery and wonder—players like Liszt or Paganini were often thought to possess superhuman or even supernatural talent—mid- to late nineteenth-century scientists brought musical performance back down to earth.[33] Victorian acousticians understood music making as a fundamentally physiological act, reliant upon the sophisticated training of muscles and nerves and the recruitment of appropriate amounts of strength and energy necessary to produce waves of sound rapid and frequent enough to turn into music. Helmholtz, for instance, wrote of the musician's ability to feel (due to the "musical vibrations of solid bodies") the "trembling of the reed in the mouthpiece of a clarinet, oboe, or bassoon, or of his own lips in the mouthpieces of trumpets and trombones."[34] Music making was a process dependent upon intense, habitual, bodily contact between players and instruments. As the German pianist Xaver Scharwenka wrote in 1895, "The fingers must *feel* the keys."[35] The mid- to late nineteenth century witnessed a proliferation of performance manuals and technical guides that detailed the physical movements necessary for musicians to play most effectively—whether to strengthen the "extensor and flexor muscles" of their fingers to improve agility, to strive for "nimbleness and flexibility of limb," or to "pay special attention to the position of the hand, the elasticity of the joints, and the sources of strength brought into play in the movements of hand and arm."[36] For E. S. J. van der Straeten, the misogynist cello instructor discussed in chapter 2, cello technique stood "in closest relationship to athletic achievements, and must therefore be acquired in precisely the same manner, *i.e.* by gymnastic exercise, patient practise and careful training."[37]

The Victorian musical marketplace capitalized on this new emphasis on "athletic" training for pianists and string players. Music periodicals advertised scores of exercises and devices that promised to strengthen musicians' wrists, fingers, and arms.[38] For instance, an 1889 letter in the *Musical Herald*'s "Correspondence" section noted that the playing of pianists with "defective"

154 | Sounding Bodies

fingers could be improved by "a course of finger gymnastics, prepared by one who has a thorough knowledge of the anatomy of the hand, as well as a practical experience of the requirements of piano *technique*."[39] In February 1888, the Royal College of Organists hosted a demonstration of several new devices invented by the musician W. Macdonald Smith, all of which were geared toward the "Physiology of Pianoforte Playing."[40] The dactylergon—a "complete gymnasium for pianists," as it was advertised—was designed to help players develop strength and dexterity in their finger muscles.[41] The technicon (see fig. 6.1) was an adjustable mechanism that attached to the

Figure 6.1. Advertisement for the technicon, c. 1887, from the *Monthly Musical Record* (May 1, 1887): 119. *Source:* Google Books. Public domain.

piano and placed weight on the pianist's hands in order to train their finger muscles—"reducing," one 1887 advertisement read, "the physiological side of pianoforte playing to a systematic and intelligible basis."[42] As James Q. Davies writes, such devices would permanently alter the performers' bodies in service of their instruments: "Instead of 'forming the hand,' virtuosos were drawn to 'deform' or better 'transform' the hand. . . . Pianistic hands were increasingly bound, clamped, or levered into extraordinary positions so as to achieve a permanent state of transfiguration."[43] In mid- to late nineteenth-century performance culture, then, players did not need to "command" their instruments so much as mold their bodies to them.

For wind and brass musicians, instrument playing involved not only the fingers but also the breath, lungs, and muscles of the face, chest, and stomach. An 1881 article in the *Monthly Musical Record* recounted a speech by W. H. Stone for the Musical Association in which he discussed how tuning wind instruments involved adjusting one's lips and mouth: "Flutes were commonly sharp, and the notes were so easy to modify in this respect by turning the embouchure inwards or outwards from the lip. . . . The oboe [was] . . . easy to flatten as well as to sharpen, from its thin double reed, indeed it often sank after heavy playing from the failure of the player's lip-muscles."[44] An 1854 letter to the editor of the *Musical World* about the "importance of the nose" in music making similarly described the intricate corporeal processes involved in woodwind performance:

> In practising the flute, oboe, clarionet, and bassoon, sustained notes, besides opening the chest and giving a *freer* respiration than would result from any other method of practice, forms the *embouchure*, by drawing into action the muscles of the neck and face, which cause a pressure of the interior or soft parts of the mouth and lips, and the fingers, by *remaining* firmly pressed in the different positions, gain an elasticity which *insures* the exccution, and gives an energetic and steady hold of the instrument.[45]

Playing a wind instrument was, above all, a visceral process—one that involved intensive breathwork, energy output, and muscular contortions. A successful performance could be *"insure[d]"* by attending closely to the body's respiratory and muscular systems.

Victorian scientists and music practitioners considered the bodily processes involved in music making, but they also emphasized the corporeal lives of instruments themselves. While instruments have been anthropomorphized for centuries—the parts of a violin have long included the "head," "neck," and "belly"—nineteenth-century musical thinkers, drawing on contemporary acoustics and physiology, described instruments as *actually* alive, as embodied entities with their own "nerves" and "muscles" and capacities for sentience and action.[46] An 1895 article in *The Strad* titled "Nerves of the Violin" suggested, "Of all the musical instruments known to man the violin possesses the most nervous temperament and the most delicate organism."[47] The writer argued that violins were "creatures" with "bodies"—every violin had a "beautifully curved back and belly" as well as "ribs"—that were sensitive to human touch:

> Violins have temperaments. Each instrument has its characteristics, peculiarities and weaknesses. . . . [The violin] demands as much care and attention as an invalid. If it is exposed to damp air, many hours' playing are required to coax back its tone, and if it gets chilled it grows as husky and as harsh in voice as any orator with a delicate throat. It feels every change of wind and temperature, and a man who knows his own violin sufficiently well can tell by drawing the bow across the four strings whether there is prevailing outside an east wind laden with moisture or a dry, cool wind from the north or west.[48]

This writer does not simply anthropomorphize the violin but views it as a body in its own right, physically formed for the emission of sound. The references to the "back" and "belly" are not metaphorical but rather indicative of actual physical features that have real acoustical purposes, as the "curves" of the violin facilitate particular sonic emissions. The violin's body is dynamic and variable, sensitive to the physical properties of air and climate. While, as mentioned in chapter 1, many eighteenth- and nineteenth-century philosophers and anatomists used string instruments as analogies for the human nervous system, here violins are described as having their *own* nerves. The violin is a quasi-living being that physically responds to the atmosphere around it, adapts its "voice" in turn, and possesses its own "strain" and control over the sound. The interaction between player and violin, then, is one of mutual sensitivity and response.

As the Victorian musical world insisted on the embodiment of both humans and instruments, Victorian writers located new opportunities to describe musicians and musical objects as feeling, sensing, and acting—and, crucially, *inter*acting—beings. If both players and instruments possess sentient bodies, then the interactions between them had the potential to be profoundly—and mutually—intimate. Acoustical science thus propelled even the most canonical Victorian authors to conceive of transgressive forms of queer kinship.

Sympathetic Kinship: George Eliot's *The Mill on the Floss*

As chapter 4 discussed, Eliot frequently imagined the kinds of human interactions that music could facilitate. She was also fascinated by the materiality of musical instruments themselves. In her 1873 poem "Stradivarius," for instance, Eliot celebrates the violin maker Antonio Stradivari as a humble, hardworking man whose craftsman skills are endowed by "God Himself"—certainly a nod to her broader ethical ideals and her commitment to those who "rest in unvisited tombs"—but also attends to the mechanical construction of musical instruments and their interactions with their human players.[49] Stradivari's violins are constructed of the "finest maple" that "serves / More cunningly than throats, for harmony."[50] Eliot foregrounds the finely crafted and responsive instrument as crucial to the virtuoso's successful performance:

> Your soul was lifted by the wings to-day
> Hearing the master of the violin
> [. . .]
> [. . .] but did you think
> Of old Antonio Stradivari?—him
> Who a good century and half ago
> Put his true work in that brown instrument,
> And by the nice adjustment of its frame
> Gave it responsive life, continuous
> With the master's finger-tips and perfected
> Like them by delicate rectitude of use.[51]

Here, Eliot demonstrates her awareness of music making as a collaborative, mutual process—not only between human instrument-maker and player but also between player and instrument itself, which possesses its own body ("frame"), animacy ("responsive life"), and capacity for development ("perfected / . . . by delicate rectitude of use").

While "Stradivarius" celebrates the material ties between craftsman, player, and violin, *The Mill on the Floss* foregrounds human-instrument intimacy as a crucial source of relief from a life filled with fraught human relationships. Although many critics, including myself in chapter 4, focus on Maggie mainly as a music *listener*, here I examine the rare moments in which Maggie appears as a music *player* in her own right. She may vibrate to Stephen's voice, yet she also enjoys private and profound moments of consolation as she plays the piano herself. Music not only serves as a medium for Maggie's relationships with other humans, but it also provides her with another, crucial source of physical intimacy altogether. It is with her piano, rather than any of her human acquaintances, that Maggie enjoys her most gratifying erotic relationship.

Maggie first begins to cultivate her intimacy with the piano when she goes to live with Lucy Deane at St. Ogg's and finally has the leisure time to play. Lucy, aware of Maggie's passion for music, tells her that she wants her to have "quite a riotous feast of it"—a phrase that reflects Lucy's recognition of music's vigorous and nourishing potential for Maggie.[52] Overjoyed at Lucy's offer, Maggie recalls a time when she worked for a family in Laceham and got to play their piano: "You would have laughed to see me playing the little girls' tunes over and over to them, when I took them to practice . . . just for the sake of fingering the dear keys again."[53] In this aside, Maggie does not simply express a vague fondness for the "dear" piano but rather recalls a vivid tactile experience. She remembers not the social atmosphere of the performance, nor what she played, but the feeling of her fingers on the keys—a feeling she is eager to recreate.

At St. Ogg's, Maggie does indeed make a "riotous feast" of her piano playing:

> It was pleasant, too, when Stephen and Lucy were gone out riding, to sit down at the piano alone, and find that the old fitness between her fingers and the keys remained and revived, like a sympathetic kinship not to be worn out by separation—to get the tunes she had heard the evening before and repeat them again and again until she had found a way of producing them so as to make them a more pregnant, passionate language to her. The mere concord of octaves was a delight to Maggie, and she would often take up a book of Studies rather than any melody, that she might taste more keenly by abstraction the more primitive sensation of intervals.[54]

In some ways, this scene casts Maggie as a music listener, one for whom music represents a "pregnant" and "passionate" language—terms that certainly evoke embodiment and desire, with "pregnant" connoting richness but also reproduction. Readers also sense that Maggie's attraction to the music is natural and instinctive (a "primitive sensation") in ways that evoke contemporary acoustical and evolutionary discussions of humans' innate predilections for music, discussed in earlier chapters.

However, Maggie's status as a player in this scene demands just as much attention as her role as a listener. In this moment, readers witness Maggie's aptitude as a performer, her inherent musical skill, and her deeply rooted passion for music. Da Sousa Correa writes that, in this scene, Maggie "modifies Stephen's songs to make them a more potent 'language' of self-expression."[55] I argue that Maggie's creativity here hinges on her physiological relationship to her music. She possesses a strong musical memory. Her ability to replicate the sounds she heard "the evening before," even when her mind is "abstract[ed]," evokes Lewes's writings about musical muscle memory, especially his notion that the "Automatic Actions" and "muscular contractions" necessary for music making could be repeated and "performed while the mind is otherwise engaged."[56] Maggie exhibits the skills to absorb the music she hears as well as the ability to recreate and manipulate it.

Yet this scene foregrounds not only Maggie's proclivities for musical creativity and memory but also her physical relationship to the piano itself. The moment hinges on the tactile exchange between Maggie and the piano keys. The phrase "old fitness" conveys Maggie's inherent musical aptitude and proclivity for "rudimentary harmony and melody," as Weliver writes, though it also indicates Maggie's visceral attachment to her instrument.[57] The evolutionary undertones of the term *fitness*—a term that suggests the survival of successful, inherited traits—hints that Maggie's haptic relationship to the piano is natural and organic. The fact that this compatibility is "old," "remain[ing]," and "not to be worn out by separation" casts Maggie's musicality as embedded in her body and innate to her physicality, improved by ("revived") but not reliant upon repeated practice. *Fitness* also extends beyond insinuations of her evolutionary aptitude for music, connoting affinity as well as adaptation. In Ahmed's terms, Maggie is "oriented" to the piano, her fingers primed and ready for music making.[58] The notion that Maggie's fingers "fit" with the keys suggests a kind of tactile sensitivity and receptivity that is natural, comfortable, and even potentially mutual (as "fitting" requires reciprocal accommodation).

Moreover, the phrase "sympathetic kinship" more explicitly captures the reciprocal intimacy between Maggie and her instrument. "Kinship" indicates that the bond between Maggie and the piano is so intense that it approaches the realm of the familial—of physical, conjugal ties or shared ancestry. In an acoustical context, the fact that this kinship is "sympathetic" further signals a mutual, responsive action between player and instrument—*sympathetic*, of course, being Helmholtz's term for the resonances between sound waves and physical objects. Maggie does not simply display a vague affection for her piano but also draws sustenance from the feelings of her fingers on the keys, which respond in sympathy to her "touch."

This scene thus offers a glimpse of another, more provocative angle of Maggie's musicality and her erotic life than critics often address. Whereas scholars like da Sousa Correa have argued that music grants Maggie moments of transcendence and escape from her social world—it "lifts her above her immediate circumstances"—I instead suggest that piano playing enables Maggie to experience an affirmative moment of closeness *with* the material world itself, with the piano that is right in front of her.[59] While human relations—particularly sexual ones, as chapter 4 showed—are, to say the least, vexed for Maggie throughout the novel, piano playing grants her rare, sustaining moments of erotic intimacy. Rather than a conduit for her relationships with other humans, then, the piano represents an important outlet for—and an animated respondent to—Maggie's touch. It is the piano, not her brother, suitors, or community members, that serves as Maggie's most "fit" companion. Maggie must look to the material world—the world of physical objects that can sound and vibrate under her fingers—to gain access to her most essential bodily sensations.

Cello Friendship: Anthony Trollope's *The Warden*

Unlike Eliot, Trollope is rarely discussed in a musical context. Music seemed to play almost no role in his life; his letters and biographies seldom mention concerts or songs, and players and instruments appear only sporadically throughout his wide oeuvre of fiction.[60] Yet in *The Warden*, the first novel in the Barsetshire series, Trollope grants a small but meaningful role to a violoncello, who is the protagonist Septimus Harding's loyal "friend" and his main source of bodily and emotional sustenance.[61]

At first, music seems merely to symbolize Septimus's troubled social and political relationships. The precentor of Barchester Cathedral and warden of the almshouse Hiram's Hospital, Septimus has his position threatened when the young reformer John Bold accuses him of corrupt business practices. These incidents imperil Septimus's economic and family life, particularly his relationship with his daughter Susan, as well as his standing in the

Barchester community. His social alienation is reflected in others' lack of appreciation for his music. An amateur violoncello player and author of an expensive volume of church music, Septimus plays "daily to such audiences as he could collect, or, *faute de mieux*, to no audience at all."[62] In a later scene, Septimus sits in a garden and plays from a "much-laboured and much-loved volume of church music," but his listeners are only interested in "appearing" to enjoy the performance, and two men even refuse to attend because Septimus's music is not "to their taste."[63]

Though he lacks appreciative human listeners, Septimus finds a crucial source of fellowship in his cello, "that friend of friends, that choice ally that had never deserted him, that eloquent companion that would always, when asked, discourse such pleasant music."[64] The cello serves not as a meager replacement for the human relations Septimus lacks but as a superlative and steadfast companion. The reference to the cello's "discourse" does not merely personify the instrument but captures a literal act of mutual sonic exchange, as the cello can actually understand Septimus's requests ("when asked") and respond in kind with "pleasant music."

The novel repeatedly emphasizes the visceral contact between human and instrument. Throughout the text, we see Septimus holding the cello "between his knees," "draw[ing] his bow slowly across the plaintive wires," and "twist[ing] and retwist[ing]" the tuning pegs.[65] In one scene, he "works with both arms till he falls into a syncope of exhaustion against the wall."[66] Cello playing is a process that involves the "work" of his whole upper body, as well as such an intense output of physical energy that he nearly faints.

Counterintuitively, Septimus and his cello enjoy their deepest forms of physical intimacy when they are apart. Early in the novel, readers learn that whenever Septimus is upset or anxious, his body begins to automatically rehearse the actions of cello playing. During a particularly troubling conversation with the archdeacon about Bold's petition, Septimus begins to play upon an invisible violoncello:

> The warden still looked mutely in his face, making the slightest possible passes with an imaginary fiddle bow, and stopping, as he did so, sundry imaginary strings with the fingers of his other hand. 'Twas his constant consolation in conversational troubles. While these vexed him sorely, the passes would be short and slow, and the upper hand would not be seen to work; nay the strings on which it operated would sometimes lie concealed in the musician's pocket, and the instrument on which he played

would be beneath his chair. But as his spirit warmed to the subject . . . he would rise to a higher melody, sweep the unseen strings with a bolder hand, and swiftly fingering the cords from his neck, down along his waistcoat, and up again to his very ear, create an ecstatic strain of perfect music, audible to himself and to St Cecilia, and not without effect.[67]

Some scholars read Septimus's pretend playing as an odd, "trance-like," antisocial spectacle that renders him a kind of "caricature."[68] "From the point of view of the realist," Jarlath Killeen writes, "Harding's wild gesticulations . . . look[] like a kind of psychosis."[69] Readers might also interpret his imaginary cello playing as an "abstract and hollow—or, indeed imaginary" search for gratification.[70] Yet, in the context of Lewes's and Spencer's discussions of instrument playing as something that could occur "while the mind is otherwise engaged," "while conversing with those around," or "while [the] memory is occupied with quite other ideas," as discussed in chapter 2, Septimus's "wild gesticulations" emerge not as symptoms of derangement but as evidence of the embeddedness of his musicality in his body.[71] For Septimus, as for Valérie, the actions of music making are so deeply embedded in his body that he is able to automatically reproduce them even in non-musical contexts. Even though Septimus is not physically holding the cello, he is able to imagine and to physically *feel* his hands moving over the strings. This process is one that involves dynamic physical exertion, as Septimus uses his whole body (fingers, hands, neck, torso, and ear) to play, and intimate bodily contact (even imagined) with the strings and cords of the cello. Septimus can even hear the rapturous music he and the cello would co-produce as an "ecstatic strain"—a phrase that connotes a sense of otherworldly transcendence (particularly with the reference to St. Cecilia) and also more material sensations of bodily tension, yearning, and pleasure. In this context, Septimus's passes, sweepings, and fingerings appear not as the comic gesticulations of a "psychotic" man but rather as a set of natural, automatic bodily actions explained by the physiological processes of muscle memory. Septimus's relationship to the cello is so intense that his musicality has embedded itself in his brain, muscles, and nerves in the way Lewes and other scientists have described.

Septimus draws on his musical muscle memory at several other points throughout the novel. During a particularly vexed negotiation in the attorney general's chambers, for instance, Septimus "plays" with vigorous intensity:

> And, as he finished what he had to say, he played up such a tune as never before had graced the chambers of any attorney-general. He was standing up, gallantly fronting Sir Abraham, and his right arm passed with bold and rapid sweeps before him, as though he were embracing some huge instrument, which allowed him to stand thus erect; and with the fingers of his left hand he stopped, with preternatural velocity, a multitude of strings, which ranged from the top of his collar to the bottom of the lappet of his coat. Sir Abraham listened and looked in wonder. As he had never before seen Mr Harding, the meaning of these wild gesticulations was lost upon him; but he perceived that the gentleman who had a few minutes since been so subdued as to be unable to speak without hesitation, was now impassioned,—nay, almost violent.[72]

Septimus's body is so well trained in the precise physical acts of cello playing that he can retrieve them whenever necessary, even without thinking. Critics have identified music's healing effects for Septimus in this scene; music "empowers him to redeem his life" and propels him to take the "decisive, heroic step of dismissing his lawyer."[73] Yet the cello not only allows Septimus to retreat from or cope with his social world but also provides him with a new kind of social world altogether, one that he physically "embrac[es]" and that literally upholds him (it "allow[s] him to stand . . . erect"). It is not even entirely clear where Septimus's body begins and his cello's body ends; his clothes become the strings on which he plays "from the top . . . to the bottom." Septimus's body is so tied to his cello that it does not even need to be physically present for him to touch it.

In *The Warden*, then, Trollope foregrounds the erotic possibilities that "queer orientations" between humans and instruments can provide. Though Septimus enjoys little in the way of filial or romantic attachments, he finds a meaningful kinship relation with his cello. In a world of hostile human relations, Septimus has another companion literally at his fingertips.

Passioned Pulsings: Thomas Hardy's "Haunting Fingers"

Hardy was well aware of the physicality of music making. In his early folk novels *Under the Greenwood Tree* (1872) and *Far from the Madding Crowd*

(1874), he depicts players who "frantic[ally] bow," "saw madly at the strings," and have "perspiration streaming down [their] cheeks."[74] The tambourine players in *Far from the Madding Crowd* execute "the proper convulsions [and] spasms . . . necessary in exhibiting the tones of [the instrument] in their highest perfection."[75] Hardy recognized that good musical sound was a product of well-executed bodily convulsions. Almost half a century later, however, he wrote a poem—"Haunting Fingers"—that imagined the visceral process of music making from the perspective of the instruments themselves. "Haunting Fingers" invites readers to ponder: What would it feel like not to play but to be played *upon*, and to reverberate in turn?

"Haunting Fingers" consists of a conversation among old instruments in a museum who bemoan their obsolescence due to their owners' deaths. Far from Hardy's only lament about abandoned instruments and lost musical traditions—several of his poems, including "The String Band Dissolved" (1871), "The House of Hospitalities" (1909), "To My Father's Violin" (1916), "Old Furniture" (1917), and "Ten Years Since" (1922), feature unplayed, worm-eaten musical objects—"Haunting Fingers" is Hardy's only work written from the perspective of the instruments.[76] Some critics interpret "Haunting Fingers" as a solemn evocation of bygone cultural traditions preserved through memory, while others read the poem as a more tragic meditation on loss and grief.[77] Elaine Auyoung argues that "Haunting Fingers" "mourns the loss not only of musicians and dancers who were once at the height of human vitality, but also of specific social and aesthetic experiences that are no longer available."[78] Indeed, the "trapped" and "doomed" instruments lament their "voiceless, crippled, corpselike state" and even wish to "die" by "f[alling] to fragments" in the damp that is slowly undoing their "glossy gluey make."[79]

My interest lies less in the poem's negotiations of memory and loss than in its portrayal of the intimate, collaborative act of music making. However long ago the performances occurred, however well preserved they are in the instruments' haptic memories, however inaccessible they might now be, the poem vividly captures the erotics of musical performance and, crucially, does so from the perspectives of the instruments themselves. The musical haptics in "Haunting Fingers"—the illustrations of sensual contact between humans and players—reflect Hardy's imagination of queer, erotic formations that exist beyond the human world.[80] While other scholars have located homoerotic intimations in texts such as *Jude the Obscure* (1895) and traced Hardy's posthuman and environmental ethics (particularly in his depictions of the natural world), I propose that Hardy also foregrounds queer social formations in his imagination of musical objects that come alive at night.[81]

Hardy's poem leaves little room for anthropocentric notions of human mastery over musical instruments. Not only are the instruments the speakers of the poem, but they are also agentic participants in their music-making activities. In one moment, for instance, the clavier reminisces about the sounds they co-created with a former player: "Tones of hers fell forth with mine, / In sowings of sound so sweet no lover could withstand!"[82] Music making is a reciprocal act between player and instrument that approximates the "sweetness" of an act of love (and might even tempt or blissfully torture a "lover"). The contra-basso describes how he and his player used to "thrill / the populace through and through, / Wake them to passioned pulsings past their will."[83] The instruments remember the acoustic force of the tones they produced in tandem with their owners ("sowings," "pulsings," and "thrills").

Much of "Haunting Fingers" is composed of the instruments' lengthy descriptions of their past and present bodily sensations. Like the author of "Nerves of the Violin," Hardy grants his instruments their own "muscles," "lips," "voices," and "contours." Many lines concern the instruments' remembrances of tactile encounters with their former players. The "aged viol," for instance, "feels apt touches on him / From those that pressed him then."[84] The present tense here is certainly important, as it suggests that the player's touch (or players' touches, given the vague plurality of the referent) has impressed itself enduringly on the instrument's body. Most of the instruments recall similar instances of haptic exchange. The speaker writes, "And they felt past handlers clutch them, / . . . Old players' dead fingers touch them."[85] "Clutch" here connotes a desperate, eager holding; as Cox writes about a different nineteenth-century "clutching" scene, "The pressure inherent in 'clutching' suggests urgency, intensity."[86]

The instruments' sensations also extend beyond tactile realms: "And they felt old muscles travel / Over their tense contours."[87] These lines not only suggest that the instruments possess bodies of their own (with the word "contours" suggesting the shape or outline of bodily curves) and capacities for muscular pressure ("tense") but also depict the rubbing of (instrument) muscle up against (human) muscle. The vague referents ("they," "muscles") here further blur the boundaries between instrument and musician; the lines focus not on which body is which but on the friction between the two.

Soon, readers learn that years of musical performance have left physical traces on the instruments' bodies. The clavier's keys are described as being "filmed with fingers"—a detail that many critics point to as further evidence of the poem's interest in preservation and memory.[88] While Ahmed discusses how objects leave "their impressions on the skin surface," here, the *instruments* are the beings with "skin" that bears the physical impressions of human-object relations.[89] The "film" reflects more than traces of past touch, however. If we consider what kinds of substances might compose this "film" (oil? sweat?), the visceral possibilities further expand. Instrument playing emerges as a process of fluid exchange, as the performer leaves their finger oils or palm sweat on the keys of the instrument, which then incorporates the substance into its own body. It is difficult to imagine a more erotic encounter.

"Haunting Fingers" thus offers a vivid glimpse into Hardy's queer imagination. The social formations he depicts in the poem go beyond sexual, familial, and human realms—and even include the phenomenological experiences and subjectivities of nonhuman objects. While Hardy's humans cannot seem to escape the "catastrophic" consequences of sex, his nonhuman objects enjoy deeply erotic lives.[90] It might be, ironically, that readers must look to dead instruments in a dusty museum to find Hardy's most vibrant examples of generative social relations.

Self-Playing Instruments

Agentic, nonhuman instruments would figure even more prominently in modernist literature, as writers became preoccupied with new sound technologies and "self-playing" musical devices. Thomas Edison's invention of the phonograph in 1877 coincided with the proliferation of other instruments, such as the gramophone and the pianola (or player piano)—none of which required extensive human involvement in order to sound. Such technological musical devices pop up all over modernist literature, including in E. M. Forster's *Maurice* (gramophone), George Bernard Shaw's *Pygmalion* (phonograph), and Joseph Conrad's *The Secret Agent* and James Joyce's *Ulysses* (pianola), in passages rife with reminders that "objects, too, have agency."[91]

Yet, as this chapter has shown, Victorian writers conceived of the life force latent even in centuries-old musical objects. While the pianos, cellos, claviers, basses, and horns discussed here all need humans to sound, the humans need them, too. Writing before the explosion of self-playing musical technologies, but after acoustical scientists began to highlight music making as a fundamentally embodied activity for both player and instrument, Dickens,

Eliot, Trollope, and Hardy were uniquely positioned to imagine the kinds of queer, erotic player-instrument intimacies that were so deeply fulfilling to their characters (both human and nonhuman). Instruments cannot play without humans, and humans cannot play without instruments—nor would they want to.

Chapter Seven

Musical Hauntings and Otherworldly Erotics in *The Lost Stradivarius* and "A Wicked Voice"

In John Meade Falkner's *The Lost Stradivarius* (1895), the protagonist, an Oxford student and violinist named John Maltravers, is joined by a mysterious musical accompanist: a fiddling ghost. John's sister, who overhears the spectral duet, remarks, "It seemed to me that he was playing with a sonorous strength greater than I had thought possible for a single violin. There came from his instrument such a volume and torrent of melody as to fill the gallery so full, as it were, of sound that it throbbed and vibrated again."[1] Readers soon learn that John's doubly resonant music results not from his own "sonorous strength" but from the haunting tones of the violin's former owner, an eighteenth-century Oxford student named Adrian Temple.

Musical ghosts make several appearances in fin de siècle and early twentieth-century supernatural literature. In Richard Marsh's "The Violin" (1900), for instance, the ghost of a violinist named Philip Coursault haunts an upper-class British household. In a memorable scene, the servants and residents enter a room only to find Philip's fiddle suspended in midair, a floating bow being drawn across the strings by an invisible hand.[2] In E. M. Forster's "Dr. Woolacott" (1927), an invalid named Clesant is visited by the ghost of a farmhand, and music—"gay, grave, and passionate"—echoes throughout the house during their interactions.[3] Such literary pairings of music and ghosts are perhaps not surprising, given their conceptual similarities. After all, both specters and sounds are in many ways invisible, intangible, and ephemeral. As David Toop argues, sound is always already "a haunting, a ghost, a presence whose location in space is ambiguous and whose existence in time is transitory. The intangibility of sound is uncanny—a phenomenal presence both in the head, at its point of source and all around."[4]

Literary critics most often focus on the "intangible" status of ghosts in Victorian literature. The ghost story as a distinct genre surged in popularity in the mid- to late nineteenth century, when events such as the Fox sisters' séances in the 1840s and Dickens's publications of supernatural Christmas tales sparked a growing public interest in spiritualism, mesmerism, mediumship, and the supernatural.[5] Following the repeal of the newspaper stamp tax in 1855 and the resultant expansion of the periodical press, ghost stories proliferated throughout publications from *Household Worlds* to *The Cornhill Magazine*, which were eager to attract "all classes of readers . . . addicted to the thrill of momentarily losing rational control over the ordered Victorian world."[6] Critics such as Julian Wolfreys, Michael Cox, and R. A. Gilbert argue that ghost stories were distinct from Gothic literature of the same period, as the latter was most often populated by *hyper*-embodied figures such as vampires, devils, goblins, and monsters.[7] Victorian ghost stories, on the other hand, "spectraliz[ed]" the gothic and "dematerializ[ed]" their villains.[8] Indeed, the works anthologized in the *Oxford Book of Victorian Ghost Stories* traffic in the language of incorporeality, featuring terms like "spirit wholly immaterialized," "disembodied spirits," and " 'bodiless dead.' "[9]

Yet ghosts—like music—are not always so disembodied. Psychoanalytic and deconstructive theorists often describe ghosts as occupying a strange, liminal space. For Peter Buse and Andrew Stott, ghosts are "a stock-in-trade of the Derridean enterprise, standing in defiance of binary oppositions such as presence and absence, body and spirit, past and present, life and death."[10] As Wolfreys writes, "Neither material nor non-material, the haunting figure uncannily traverses between matter and the abstract, between the corporeal and the incorporeal, incorporating itself within both, while never being available corporeally."[11] Sonically minded readers may notice that "ghost" could be replaced with "music" in either of these formulations. Like Wolfreys's "haunting figure," music straddles the abstract and the material, the absent and the present, the disembodied and the embodied.

This chapter focuses on two supernatural tales—Falkner's *The Lost Stradivarius* and Vernon Lee's short story "A Wicked Voice" (1890)—that use contemporary acoustical science to attribute material qualities to (allegedly) immaterial ghosts. Both *The Lost Stradivarius* and "A Wicked Voice" depict human protagonists who are haunted by the music of long-dead performers from previous centuries. John is visited by Adrian's ghostly music, while Lee's protagonist Magnus is tantalized by the spectral sounds of an eighteenth-century castrato opera singer named Zaffirino. Oddly, though, both Falkner and Lee imbue their patently non-realist literary works with the empirical language of acoustical science, a field of study with which both were familiar, as I will discuss. Through references to vibrating bodies, objects, and spaces, Falkner and Lee are able to bring their "bodiless" ghosts firmly into the physical world.

Acoustical science allows Falkner and Lee to imagine musical hauntings as deeply erotic encounters. As Falkner's passage above demonstrates, the language of acoustics—the music of "sonorous strength" that "throb[s] and vibrat[es]" the walls—grants Adrian's ghost a palpable sonic presence in the "real world." While Wolfreys suggests that ghosts are never "available corporeally," the ghostly music in *The Lost Stradivarius* and "A Wicked Voice" vibrates chairs, shakes walls, penetrates the atmospheres of rooms and cities, and incites tangible pleasures in listeners' bodies.[12] Though John, Magnus, and their fellow listeners cannot always *see* the ghosts that haunt them, they can *hear* them—and thus, in the context of nineteenth-century acoustical science, *feel* them.

The musical hauntings in these works are queer and erotic in ways that resemble the encounters discussed in earlier chapters, but they also manifest additional concerns, as they unfold beyond the limits of the physical world and occur outside of normative arrangements of space and time. First, like the authors discussed in chapters 5 and 6, Falkner and Lee foreground erotic intimacies that depart from the realm of cross-sex, reproductive relations. Both Adrian and Zaffirino are largely coded as male figures, and thus their exchanges with the male protagonists are often read as evocations of same-sex desire, as I will discuss further. However, these works also depict more multifaceted forms of queerness. Zaffirino, for instance, is a castrato, an opera singer who was castrated before puberty to preserve his high singing voice, lung capacity, and vocal flexibility. His status as an ambiguously sexed being whose voice and body reflect, as Roger Freitas writes of castrati, "the erotic mixture of masculine and feminine qualities," renders the erotic relations between him and Magnus even more indeterminate.[13] In addition, in an echo of the human-instrument intimacies discussed in chapter 6, John is just as attached to Adrian's violin as he is to the specter of Adrian himself.[14]

Moreover, by "spectraliz[ing]" the body of the performer, Falkner and Lee are able to draw attention to the erotic powers of music itself. While several of the works discussed in previous chapters depict humans' erotic relations to musical notes and vibrations, never is the erotic force of music clearer than when the performer and instrument are not-quite-physically present. Both Falkner's and Lee's protagonists yearn not for the bodies of the ghostly players, but for the notes, melodies, and sonic vibrations of their music; John and his companions long for the tones of the *Gagliarda* suite, and Magnus wishes to "mingle" with and "suck in" Zaffirino's eighteenth-century vocal airs.[15] The queerness of these musical moments, then, arises from their intermingling of the sensations aroused by player, listener, and music. By making audition, rather than overt tactility, the primary source of intimate exchange, Falkner and Lee present broad understandings of what constitutes erotic interaction altogether and where one might find it.

In addition, the musical hauntings in Falkner and Lee are queer and erotic because they are, after all, *hauntings*. From Terry Castle's "apparitional lesbian" to Diana Fuss's discussion of ghostly elegies, queer theorists have for decades described ghosts as inherently queer figures because they "traverse categorical distinctions," as Muñoz writes, and exist outside of norms of humanness, race, gender, and sexuality.[16] As ghosts, Adrian and Zaffirino are not-quite-human and not-quite-embodied. While chapter 6 also explored human-nonhuman intimacies, this chapter further surpasses the human realm by delving into the world of the (un)dead. Dickens and his contemporaries were eager to grant embodied life to physical objects (musical instruments), whereas Falkner and Lee focus on figures who do not have bodies at all. The musical hauntings in *The Lost Stradivarius* and "A Wicked Voice" suggest that erotic encounters can occur beyond the land of the living, including with "vaporous," "fluid," and otherwise evanescent beings.[17]

Finally, as ghosts, Adrian and Zaffirino are also, by definition, figures from different times and places, occupying "ambiguous state[s] of being, both present and not, past and not," writes Carla Freccero.[18] Both Adrian and Zaffirino were inhabitants of eighteenth-century Italy (Adrian attended Oxford but lived in Naples for much of his life), and their instruments and music also represent relics from past centuries. Adrian's violin, covered in cobwebs when John finds it, was made in 1704, and the "Gagliarda" suite he plays was composed in 1744. Zaffirino haunts Magnus with old "singing-exercises," "scraps of forgotten eighteenth-century airs," and songs from "musty music-books of a century ago."[19] Certainly, from the perspectives of twenty-first-century theories of queer antisociality and queer temporality, the musical hauntings in Falkner and Lee can be said to halt the protagonists' ties to "reproductive futurism" and thwart their "chrononormative" life paths, disrupting their careers, marriages, and even life spans.[20] However, I am more interested in the ways in which these musical hauntings highlight the queer erotics of contact between the past and the present. While many critics interpret ghosts as symbols of queer degeneration and backwardness and agents of ruinous slippages back into the past (indeed, John and Magnus at times experience the ghostly music as frustrating, torturous, and damaging), I argue that the musical hauntings in *The Lost Stradivarius* and "A Wicked Voice" also offer the protagonists what Freeman refers to as "erotic contact with the past"—encounters with history that are intimate and sensual rather than objective or distant.[21] As Freeman writes of *Frankenstein*, Gothic and ghost stories "stag[e] the very queer possibility that encounters with history are bodily encounters, and even that they have a revivifying and pleasurable effect."[22] By imagining ghosts who produce music from their own times—and listeners who are able to hear, nay physically *feel*, its effects—Falkner and Lee examine the pleasurable physical sensations possible when different temporalities collide. Acoustical science enables Falkner and Lee to imagine a new way of realizing what Carolyn Dinshaw describes as the "queer . . . desire for bodies to touch across time."[23] The texts discussed here, then, are in many ways the queerest of all those discussed in this book, as they unsettle normative conceptions not only of gender, sex, and humanism but also of time and space, redrawing the boundaries of sentient being and conceiving of new modes of relation between the past and the present.

This chapter thus sheds a new, more reparative light on the queer potential of Victorian ghost stories. Many critics read fin de siècle supernatural tales as evocations of queer repression, absence, or loss, in which ghosts serve, to quote Stephen Best and Sharon Marcus, as "surface signs of the deep truth of a homosexuality that cannot be overtly depicted."[24] Critics such as Sylvia Miezkowski, Ruth Bienstock Anolik, and Cox and Gilbert argue that ghost stories repress queer desire by abstracting it into the realm of the supernatural.[25] Yet the presence of music and acoustics in *The Lost Stradivarius* and "A Wicked Voice" complicates these critical frameworks. Acoustical science enabled Falkner and Lee to render their ghosts *hyper*-present in the material world, taking on flesh and body and producing sounds that vibrate rooms and make human listeners throb and moan in response. Falkner and Lee drew on acoustical science not only to unsettle cross-sex desire or interrogate anthropocentric forms of intimacy but also to fundamentally rethink the many ways that a multitude of beings—broadly conceived—can interact and what kinds of alternative life experiences those interactions might allow.

"Acoustical Affinities": *The Lost Stradivarius*

John Meade Falkner was perhaps an unlikely author of a queer, erotic ghost story. Much of his life bore the markings of a normative, middle-class English existence. The son of a Wiltshire clergyman, Falkner attended Oxford before working his way up in the arms company Armstrong Whitworth.[26] Yet Falkner also led what Nicholas Daly refers to as a "peculiarly modern double life."[27] He worked as an honorary librarian and reader in paleography at the University of Durham and was an avid collector of medieval manuscripts.[28] He also published three novels: *The Lost Stradivarius*, *Moonfleet* (1898), and *The Nebuly Coat* (1903).

In many ways, *The Lost Stradivarius* reads as a work that attempts to punish and erase sexual deviance. The novel is narrated by John's sister Sophia and also includes a lengthy reminiscence by John's Oxford friend William Gaskell. Their narrations, framed as addresses to John's surviving son Edward, lament John's tragic fate, which they attribute to his "actions . . . which may not seem becoming to a noble gentleman, as he surely was."[29] In a proclamation representative of the moralistic tone of her narrative, Sophia insists, "We must humbly remember that to God alone belongs judgment, and that it is not for poor mortals to decide what is right or wrong in certain instances for their fellows, but that each should strive most earnestly to do his own duty."[30] Though the friendship between William and John is sometimes vaguely coded as sexual—we learn that the two men would sit up "without lights until the night was late" and that these musical evenings gave both men "much gratification"—most of William's narration is composed of his critique of John's musical affinities.[31]

Both Sophia and William describe Adrian's haunting music as the source of John's physical, mental, and moral decline. They intimate that, back in the eighteenth century, Adrian engaged in licentious affairs with both women and men and lived a "notoriously evil life."[32] Music was Adrian's main mode of seduction and debauchery, "put by him to the basest of uses."[33] In terms that echo Victorian sexological discussions that linked a proclivity for music to homosexuality or effeminacy, Sophia and William describe John as particularly susceptible to Adrian's ghostly musical influence.[34] Sophia indicates that John had always "exhibited some symptoms of delicacy," and William notes that he was, "like most cultured persons—and especially musicians,—highly strung and excitable."[35] As John becomes enamored with Adrian's ghostly music, Sophia begins to see the same metaphorical "mark of the beast" upon him.[36] According to Sophia and William, John's repeated ghostly encounters ruin his gentlemanly existence and threaten his religious, professional, marital, and familial obligations. He abandons Christianity for the "seductive" philosophy of Neoplatonism, and he does "little work" and forgoes any pursuit of a career.[37] Though John eventually marries Constance Temple—a descendant of Adrian's whom, we learn, shares Adrian's complexion (a detail that renders the union between John and Constance a rather on-the-nose nod to John's sublimated desires)—John casts her aside in favor of playing Adrian's violin. He spends their honeymoon in Naples "devoted . . . much to the violin," often leaving Constance "alone," and he repeatedly denies her even "ordinary . . . mark[s] of affection."[38] In a thinly veiled insinuation of John's sexual denial of Constance, Sophia notes, "My brother had evidently ceased to take that pleasure in her company which might reasonably have been expected in any case under the circumstances of a recent marriage, and a thousand times more so when his wife was so loving and beautiful a creature as Constance Temple."[39] When Constance asks to accompany John on the piano, he rejects her by invoking their musical incompatibility: "I cannot keep time with you."[40] For John, then, music is far from the "humanising and educational agent" that William, echoing Victorian moralists like Haweis, believes it should be; music renders John subject to his "sensual appetites" and ruins his prospects for a successful aristocratic existence.[41] As Daly writes, it is thus "possible to read *The Lost Stradivarius* as a deeply phobic text that turns Paterian aestheticism into a sort of Hammer Horror Satanism."[42]

And yet, embedded within the novel's cautionary and moralistic framework are counternarrative moments (found mostly in fleeting observations by Sophia or William or in quotations of John's own comments) that highlight—and do not entirely indict—the erotic gratification that John experiences while playing Adrian's violin or hearing his ghostly music.[43] In many of Falkner's scenes of musical haunting, ghostly music emerges not as dangerous or wicked, nor as a depraved fantasy or figment of John's diseased imagination, but as a real, sonic event explained by acoustical science. Falkner often depicts John's responses to Adrian's music—as well as the music's effects on the physical environment and on the bodies of other listeners, including Sophia and William—in terms of vibration, resonance, and tonality. Adrian's ghostly violin playing exerts hyper-material effects on the "real" world, filling space, shaking walls, and making human bodies quiver. Despite its moralistic bent, then, *The Lost Stradivarius* also makes space for characters to experience fulfilling erotic encounters that stretch not only beyond the realm of cross-sex desire but also outside of the earthly world and back into past times and spaces.

From the beginning of the novel, Adrian's hauntings are inextricable from acoustical science. The hauntings begin when William brings some books of seventeenth-century Italian music back from a holiday in Rome. One piece, the "Areopagita" duet for violin and harpsichord by an Italian composer named Graziani, catches John's eye, and he begins to play it. When he reaches the piece's "Gagliarda" movement, John perceives a "perfectly familiar" sound, a creaking of the wicker chair in the corner of the room: "The illusion was so complete that my brother stopped playing suddenly, and turned round expecting that some late friend of his had slipped in unawares, being attracted by the sound of the violin, or that Mr. Gaskell himself had returned."[44] Yet, when John turns around, the chair is empty. He "easily" arrives at an acoustical explanation for the creaking based on "theories of vibration and affinity."[45] In a direct reference to Helmholtz's theories of sympathetic vibration, which would have been taught at institutions like Oxford during Falkner's time, John determines that "there must be in the wicker chair osiers responsive to certain notes of the violin, as panes of glass in church windows are observed to vibrate in sympathy with certain tones of the organ."[46] The next evening, when William joins John to play the piano accompaniment of the "Gagliarda," he too hears the creaking of the chair. John again surmises, "Certain parts of the wicker-work seem to be in accord with the musical notes and respond to them. . . . It is the vibration of the opening notes which affects the wicker-work."[47] As the weeks go on and the creaking continues, both John and William "professed to be quite satisfied that it was to be attributed to acoustical affinities of vibration, between the wicker-work and certain of the piano wires, and indeed this seemed the only explanation possible."[48] Nineteenth-century acoustical science provides John and William with a reasonable explanation for this mysterious sonic phenomenon.

Soon, though, the creaking of the chair begins to defy acoustical logic. One night, John tests the sympathetic vibration theory by playing "with more vigour and precision than usual."[49] As he plays, he perceives a spectral figure beginning to take shape: "It was that night that my brother, looking steadfastly at the chair, saw, or thought he saw, there some slight obscuration, some penumbra mist, or subtle vapor, which, as he gazed, seemed to struggle to take human form."[50] It is now unclear whether the creaking of the chair is caused by the responsive "osiers" sympathetically vibrating to the music or by a spectral listener perching upon it.

The slippage between the scientific and the supernatural in these scenes is illustrative. While sound science fails to fully account for the creaking (at least the third time), the invocation of Helmholtzian acoustics nonetheless highlights the ghost's palpable sonic presence—its ability to enter into and tangibly affect the physical world. Though invisible or barely visible (a "slight obscuration"), the ghost is unmistakably audible, producing sound waves that travel through the air to reach John's and William's ears and vibrate physical objects. Ghostly haunting is not some kind of demonic, spooky, or otherworldly event but something akin to (and in fact hard to disentangle from) an acoustical phenomenon.

The ghost of Adrian is musical as well as noisy, and he soon begins to accompany John's playing with his own violin tones. Shortly after the chair-creaking episodes, John finds an old violin tucked away in the cupboard of his Oxford sitting room. Like the protagonists discussed in chapter 6, John enjoys meaningful physical intimacy with the instrument even before he begins to play it: "holding it on his knees," "admiring it," cleaning it with "gentle handling," and gazing at "the delicate curves of the body and of the scroll."[51] When John begins to play the violin, the ghost of Adrian accompanies him on his own spectral instrument. One evening, John performs the "Areopagita" suite for William, who is so impressed by the music's powerful reverberation that he devises yet another acoustical explanation for it based on John's rearrangement of the furniture in the room: "The change seems to me to have affected also a marked acoustical improvement. The oak panelling now exposed on the side of the room has given a resonant property to the wall which is peculiarly responsive to the tones of your violin. While you were playing the *Gagliarda* to-night, I could almost have imagined that some one in an adjacent room was playing the same air with a *sordino*, so distinct was the echo."[52] Again, the boundary between the scientific and the supernatural is quite vague. The doubly sonorous nature of the music—the "impression of the passages being chorded, or even of another violin being played"—evokes Helmholtz's theory of compound and partial tones. Helmholtz argued that compound tones are formed by a combination of various pitches, or partial tones, which vibrate more rapidly than their "prime" tones.[53] As a result, he wrote, the human ear does not just hear one musical tone but rather "becomes aware of a whole series of higher musical tones" (partial tones).[54] Both Helmholtz and Tyndall associated partial tones in particular with the violin. Tyndall wrote, "When two violins are sounded, we have not only to take into account the consonance, or dissonance, of the fundamental tones, but also those of the higher tones of both."[55] The sonic situation that Falkner describes, then, could be explained either by the acoustical properties of the room—the sound resonating in new ways against the oak paneling—or by the presence of another violinist whose music produces an additional set of prime and partial tones. In either case, the ghostly music reflects not a diseased or degenerate aspect of John's imagination but an acoustical event—one perceived by both John and William.

These spectral duets persist as the novel continues. After John has the violin evaluated and rehabilitated by an expert in London, who deems the instrument to be a rare creation by the legendary violin maker Antonio Stradivari, John returns home for a summer holiday, eager to play. Upon hearing John perform, Sophia comments, "He had naturally expected from the instrument a very fine tone, yet its actual merits so far exceeded his anticipations as entirely to overwhelm him. The sound issued from it a volume of such depth and purity as to give an impression of the passages being chorded, or even of another violin being played at the same time."[56] This doubly resonant music is deeply pleasurable, even life-affirming, for John, who has recently been ill. Sophia remarks, "His performance was greatly improved, and . . . he was playing with a mastery and feeling of which he had never before been conscious. . . . He had actually acquired a greater freedom of wrist and fluency of expression, with which reflection he was not a little elated."[57] The musical hauntings are more invigorating than damaging, as John finds a renewed "deep passion" for music and gains "more power" in his playing than ever before.[58] Later, when William hears John play the Stradivarius, he too reveals, "As I came along I was quite spellbound by your music. I never before heard you bring from the instrument so exquisite a tone: the chorded passages were so powerful that I believed there had been another person playing with you."[59] Rather than ruinous forces that drag John into a realm of deviance and degeneration, Adrian's musical accompaniments are palpable vibrations from the past that facilitate bodily pleasure, health, and artistic expression for John and his listeners.

The ghostly-human duets exert material effects on physical spaces as well as human bodies. One night, at the Maltravers estate, Sophia and Constance are awakened by the "distant and very faint murmur . . . upon the sleeping air," which "seemed to resolve itself into the vibration, felt almost rather than heard, of some distant musical instrument. . . . All was deadly still, but I could perceive that music was being played somewhere far away."[60] Constance describes the sound as a "vibration" that is "felt almost rather than heard"—an acoustical detail punctuating the music's physical and physiological force.[61] When the two women make their way down the corridor, Sophia laments that the sounds are becoming "even more detestable to [her] ears."[62] Yet both women are drawn to the music:

> It was as if some irresistible attraction drew us towards the music. Constance took my hand in hers and we moved slowly down the passage. . . . He was playing the violin, playing with a passion and reckless energy which I had never seen, and hope never to see again. . . . It seemed to me that he was playing with a sonorous strength greater than I had thought possible for a single violin. There came from his instrument such a volume and torrent of melody as to fill the gallery so full, as it were, of sound that it throbbed and vibrated again.[63]

Though the sounds are produced in part by a supernatural entity, they nonetheless add discernible harmonic properties to John's tones; the music permeates the whole room, taking on "volume" and "sonorous strength" and filling the gallery in ways that "vibrate" and "throb" the walls. Moreover, the pleasure of this moment is undeniable. The music is not only sonically powerful ("depth and purity," "sonorous strength," "torrent of melody") but also sensually, even sexually, evocative ("throbbing," "vibrating," and "filling the room"). The fact that both Constance and Sophia are physically drawn ("irresistibly attracted") to the sound further distances this moment from John's own "diseased" brain; the music is real and undeniably pleasurable. Sophia and Constance hear the music as being "endowed with life" (despite its dead player) and find themselves wholly engrossed in the sound, "feel[ing]" the musical vibrations within their bodies. The text insists on the widespread attraction of Adrian's vibratory, ghostly music, which results not from the tortured imagination of one man's fantasies, but rather from the sound waves that are irresistible to all as human and nonhuman worlds collide. Sophia, Constance, John, Adrian, the music, and the world of eighteenth-century Italy are all drawn together in intimate proximity.

As mentioned earlier, Falkner ultimately kills off his "deviant" figures. John eventually withers away, a victim of both brain fever and, Sophia believes, his "fatal passion for the violin."[64] The ghost of Adrian disappears forever, and William and Sophia burn the violin that precipitated the entire narrative. Yet, while *The Lost Stradivarius* ultimately forecloses many of the erotic possibilities it introduced, most memorable for readers are the moments of musical pleasure that undercut the text's rather unmemorable moralistic conclusion. After all, it is during his nightly musical encounters with the ghost of Adrian, as opposed to his studies at Oxford or his passionless marriage to Constance, that John experiences the most powerful forms of bodily gratification—pleasures that are not unique to him but also experienced by Sophia, William, and Constance; by the vibrating panels in the sitting room; and by the very air that vibrates to the ghostly music.

The Lost Stradivarius thus invites readers to ponder: Is the realm of the ghostly actually more "ruinous" than the "real" world, the realm of strictly human relations that denies John so much? Are ghosts so terrifying if they allow characters access to bodily experiences that their world otherwise disavows? Which is ultimately more "haunting"—the ghostly world that enables moments of bodily stimulation and intimate encounters across time and space, or the real world that denies such experiences? Indeed, the final moment of the novel hints that a departure from the land of the living might not be the worst outcome for John. William describes the inscription on John's gravestone: "Out of this light; alas! alas! for some the light is darkness."[65] While this inscription invites many interpretations (religious, poetic, and otherwise), one reading is that it insinuates how, for John, the "light" was actually found in the darkened rooms of his Oxford dormitory, the hidden cupboard where he found the violin, and the nighttime duets with Adrian's ghost.

Moreover, in a text that so insistently blurs the past and the present, the ghostly and the human, John's death does not seem permanent or consequential. Are ghosts even that unfamiliar or distant from the human world if they are able to enter it and exert acoustical effects on its surroundings? If the past and the present, the supernatural and the natural, death and life are so muddled, then how final is John's death or the burning of the violin? If ghosts can return to the world and make music, then might John still be able to access the pleasurable sensations that music making granted him? In the world of *The Lost Stradivarius*, after all, death does not mean the end of aesthetic enjoyment; instead, death facilitates it. Though Falkner's novel may warn of the dangers of licentious transgression, it also presents possibilities for more fulfilling forms of bodily exchange, even among beings that do not have bodies at all.

The Murmur of the Castrato: "A Wicked Voice"

Like *The Lost Stradivarius*, "A Wicked Voice" can in some ways be read as a devastating tale of queer suffering, in which Magnus is the tortured victim of Zaffirino's dangerous musical influence. As in Falkner's novel, Lee's text opens with lore of Zaffirino's danger and wickedness. When Magnus is dining at his Venetian boardinghouse one evening, he overhears his fellow tenants discussing Zaffirino, otherwise known as Balthasar Cesari, an eighteenth-century castrato opera singer who wooed—and ruined—women with his singing. One man recounts how Zaffirino's singing made an aristocratic woman, the Procuratessa Vendramin, "f[a]ll into the convulsions of death."[66] Zaffirino makes Magnus, too, sometimes feel "wasted by a strange and deadly disease."[67] For Magnus, Zaffirino's music represents "corrupt and corrupting music of the Past" that repeatedly frustrates his attempts to compose a modern opera in the style of Richard Wagner, whom he hails as the "master of the Future."[68]

Indeed, most critics read "A Wicked Voice" as a tale of queer erasure, frustration, or melancholy—one that reflects biographical details about Lee's own sexuality.[69] Catherine Maxwell, for instance, proposes that the story reflects Lee's "disguised lesbianism" and her frustration at the marriage of her alleged love interest, the opera singer Mary Wakefield.[70] For Sylvia Miezkowski, the story reflects "cultural anxieties" about emasculation, the ruinous effects of male-male relationships, and the "dissolution of heteronormative structures of identity and desire."[71]

Yet "A Wicked Voice" also at times celebrates Zaffirino and Magnus's musical encounters as erotically fulfilling events. Whereas *The Lost Stradivarius* is somewhat ambivalent about the damaging effects of Adrian's ghostly visitations, Lee's text seems to relish Zaffirino's musical intrusions. As Vineta Colby and Carlo Caballero have pointed out, Lee's own musical tastes—she despised Wagner—likely rendered her eager to depict the replacement of Magnus's Wagnerian-style composition with Zaffirino's eighteenth-century Italian music, which was, according to Caballero, one of Lee's "life's great loves."[72] Throughout the text, Lee emphasizes the rapturous physical pleasures of Zaffirino's musical hauntings, which can be sexually charged but elsewhere are simply stimulating, relaxing, or invigorating. Several critics have noted that Lee's critical writings and fiction often reflect her own attempts to navigate the "contradictions of Victorian aestheticism"—the tensions between ideals of pure, aesthetic beauty and art's more organic qualities as studied by physiological scientists.[73] "A Wicked Voice," I argue, is indebted to the latter. Zaffirino's music vibrates through the air and incites intensely visceral responses in Magnus's body—the kinds of "ecstatic and even orgiastic" responses that Lee observed among her research subjects and charted in *Music and Its Lovers*.[74]

As mentioned earlier, while the erotic musical encounters in "A Wicked Voice" certainly intimate same-sex desire (both Magnus and Zaffirino are primarily referred to as male), they also go far beyond it; Zaffirino is, after all, both a castrato—a singer who embodies sexual "mutability" and "ambiguity," as Caballero writes—and a ghost, a being who is not quite embodied, not quite of this world, who straddles the past and the present, the living and the dead, the human and nonhuman realms.[75] "A Wicked Voice" is thus not a tale of repressed same-sex desire but a radical imagination of erotic intimacies that vibrate far beyond the land of the living. Queer interactions between the past and the present, the dead and the living, and the human and the nonhuman need not be "dematerialized" but can be concrete and tangible, rationally explained by the science of sound.

Zaffirino's music takes on a palpable physical presence. Magnus first hears the music when he rides on a gondola in Venice one evening, trying to "meditat[e]" on his opera:

> Suddenly there came across the lagoon, cleaving, chequering, and fretting the silence with a lacework of sound even as the moon was fretting and cleaving the water, a ripple of music, a voice breaking itself in a shower of little scales and cadences and trills. . . . A faintness overcame me, and I felt myself dissolve. . . . The murmur of a voice arose from the midst of the waters, a thread of sound slender as a moonbeam, scarce audible, but exquisite, which expanded slowly, insensibly, taking volume and body, taking flesh almost and fire, an ineffable quality, full, passionate, but veiled, as it were, in a subtle, downy wrapper. The note grew stronger and stronger, and warmer and more passionate, until it burst through that strange and charming veil, and emerged beaming, to break itself in the luminous facets of a wonderful shake, long, superb, triumphant.[76]

In some ways, the music is immaterial ("ineffable"), as readers might expect of a spectral song. Yet this passage is also rife with excessive materiality. Though Magnus initially thinks he is "going mad"—that the music might be a figment of his disordered imagination—the narrative makes clear that the music has a real physical presence. Though "scarce audible," the note develops a tangible presence, "taking volume and body" and "flesh" and filling physical space in ways that call forth Victorian acoustical scientists' notions of sound waves as series of particles that vibrate the air. The overflowing pleasures of the music are manifest in the prose itself; the run-on description of the note's accumulating qualities grants the music almost an orgasmic path—growing stronger, warmer, and more passionate and then bursting and shaking—and suggests an excess that cannot be contained. Magnus's unsettling but pleasurable physiological response ("faintness" and "dissolv[ing]") is framed as attraction not merely to another man or to an ambiguously sexed being but to music itself, to "scales and cadences and trills" that cleave and fret the Venetian night air.

As Magnus returns to his work, he feels "interrupted ever and anon" by the memory of the ghostly voice.[77] Though he feels "the need of noise, of yells and false notes, of something vulgar and hideous, to drive away that ghost-voice which was haunting me," he cannot help but think—"not without a certain pleasure"—of Zaffirino and his music, with its "voluptuous phrases and florid cadences."[78] One night, while he tries to compose, he hears a voice rising above a chorus of Venetian folk music on the street below:

> My arteries throbbed. How well I knew that voice! It was singing, as I have said, below its breath, yet none the less it suffered to fill all that reach of the canal with its strange quality of tone, exquisite, far-fetched. They were long-drawn-out notes, of intense but peculiar sweetness, a man's voice which had much of a woman's, but more even of a chorister's, but a chorister's voice without its limpidity and innocence; its youthfulness was veiled, muffled, as it were, in a sort of downy vagueness, as if a passion of tears withheld.[79]

This passage, with its list of "but" clauses, captures the indeterminate status of Zaffirino's voice; it is neither male nor female, adult nor child ("chorister's"), and it is muffled and vague. What is clear, however, is that the voice possesses the ability to travel through space ("fill all that reach of the canal") and act on human bodies. The detail of Magnus's "arteries throbb[ing]" echoes physiological studies by Hector Chomet and Alexandre Dogiel about sound's effects on the nervous and circulatory systems. Moreover, as the text makes clear, Magnus's physiological response is not his alone. As in *The Lost Stradivarius*, the protagonist is not the only one haunted by the music—a detail that further distances "A Wicked Voice" from Victorian sexology, which cast "homosexuals" as diseased or degenerate.[80] After the ghostly performance described above, Venice "echoe[s] with . . . clapping," and "every one pressed on, and clapped and vociferated."[81] For days, we learn, the "mysterious singer was the universal topic" across Venice.[82] Zaffirino's music is not a figment of Magnus's diseased imagination, nor is his physiological response unique to him; Zaffirino's music "takes flesh" among the bodies and spaces across the entire city.

The physiological pleasures of Zaffirino's music augment as the story progresses. Later, Magnus—in an attempt to escape the "malady" brought on by the ghostly music and make progress on his opera—accepts the Count Alvise's invitation to his estate in Mistrà, a villa that Zaffirino once visited. One day, Magnus enters a cathedral, where he believes a musical mass is taking place, but is soon met with Zaffirino's ghostly singing:

> And above the organ rose the notes of a voice; high, soft, enveloped in a kind of downiness, like a cloud of incense, and which ran through the mazes of a long cadence. The voice dropped into silence; with two thundering chords the organ closed in. All was silent. For a moment I stood leaning against one of the pillars of the nave: my hair was clammy, my knees sank beneath me, an enervating heat spread throughout my body; I tried to breathe more largely, to suck in the sounds with the incense-laden air. I was supremely happy, and yet as if I were dying.[83]

The passage captures music's tangibility not only through the simile ("like a cloud of incense") but also through the emphasis on music's movement through space ("ran through the mazes") and effects on Magnus's body. Magnus's sensations here—"supremely happy, and yet as if I were dying"—which many psychoanalytic or queer theorists would likely read as an instance of *jouissance* or "petit mort," are also concrete reactions of his muscular, respiratory, and nervous systems, as he sweats, grows hot, wobbles, and breathes heavily.[84] Importantly, Magnus's visceral response derives more from his contact with the music itself than with Zaffirino's body. Magnus yearns to ingest the music ("to suck in the sounds"), to merge his body with it entirely.

Magnus's musical desires are further realized as he spends more time in Mistrà. One evening, as he gazes upon the moonlit village, he hears "a note, high, vibrating, and sweet, rent the silence, which immediately closed around it. I leaned out of the window, my heart beating as though it might burst. After a brief space the silence was cloven once more by that note, as the darkness is cloven by a falling star or a firefly rising slowly like a rocket."[85] The figurative language here likens the allegedly "wicked" voice to charming natural phenomena, such as fireflies and falling stars. Yet, again, the passage goes beyond mere figurative language; it is just as preoccupied with the note's "vibrati[on]" through the air and Magnus's intense circulatory response ("heart beating as though it might burst") as it is with celestial imagery. It is in this moment that Magnus realizes that he has come to Mistrà "on purpose," in an attempt to meet the ghost of Zaffirino, "the object of [his] long and weary hopes."[86]

Magnus does finally encounter Zaffirino in quasi-physical form in the subsequent scene, in which he follows the voice through the palace and perceives the figure of "a man stooped over a harpsichord."[87] When the man begins to sing, Magnus recognizes the features of Zaffirino's voice—"that delicate, voluptuous quality, strange, exquisite, sweet beyond words, but lacking all youth and clearness."[88] He ponders, "But I recognised now what seemed to have been hidden from me till then, that this voice was what I cared most for in all the wide world."[89] As Zaffirino continues to play, Magnus fearfully realizes that unfolding before him is the vision of Zaffirino's musical murder of the Procuratessa Vendramin. Though Magnus tries to escape the room, the music continues. It is here that Magnus's musical response most closely resembles an orgasm:

> The voice wound and unwound itself in long, languishing phrases, in rich, voluptuous *rifioituras*, all fretted with tiny scales and exquisite, crisp shakes; it stopped ever and anon, swaying as if panting in languid delight. And I felt my body melt even as wax in the sunshine, and it seemed to me that I too was turning fluid and vaporous, in order to mingle with these sounds as the moonbeams mingle with the dew. . . . I heard the voice swelling, swelling, rending asunder that downy veil which wrapped it, leaping forth clear, resplendent, like the sharp and glittering blade of a knife that seemed to enter deep into my breast. Then, once more, a wail, a death-groan, and that dreadful noise, that hideous gurgle of breath strangled by a rush of blood. And then a long shake, acute, brilliant, triumphant.[90]

This passage pairs transcendent language ("moonbeams," "vaporous") with materialist imagery; the voice takes on physical qualities as it "shakes," "sway[s]," and "pant[s]." Magnus, too, experiences a rapturous physiological response; he feels his body "melt," "turning fluid and vaporous." While these words might seem to signal bodily ruin, Magnus's physical dissolution is paradoxically pleasurable, associated with sunshine and mingling sounds.

Readers might think of Magnus's "melting" as an instance of what queer theorists like Butler conceive of as "coming undone," a process that "interrogat[es] . . . the terms by which life is constrained in order to open up the possibility of different modes of living."[91] For Magnus, the sensation of bodily dissolution enables him to access pleasurable states of being ("wax in the sunshine") and new realms with which he can "mingle." The synecdochic formulation of the passage above, in which "the voice" (rather than Zaffirino himself) is the subject, highlights that Magnus's physical intimacy occurs as much with the music as with the visual sight of Zaffirino's ghost. What Magnus ultimately desires, it seems, is to depart the human world and "mingle" with the sounds he hears. The voice that unmakes him is what he "care[s] for most in all the wide world."[92] Lee thus invites readers to question: Is it possible that humans can, ironically, experience the most intense forms of bodily gratification when they depart from the strictly physical world? Can humans find their deepest experiences of physiological pleasure in encounters with figures who do not have bodies at all? Are the sounds of the past more stimulating than the "music of the future"?

While the endings of both *The Lost Stradivarius* and "A Wicked Voice" might punish sexual deviance and illustrate the fatality of same-sex desire, what remains lingering in the minds of readers is the gratification that the musical hauntings produce and the wildly expansive visions of erotic life that they enable. For John, music is the "light" in the "darkness" of his life; for Magnus, it is "what [he] cared for most in all the wide world." For Falkner's and Lee's protagonists, erotic sensations are available not in the realm of cross-sex encounters, nor even in strictly here-and-now, human spaces, but in the slippages between the past and the present, the dead and the living, the immaterial and the material. Acoustical science offers Falkner and Lee the language to imagine erotic possibilities hovering outside of the places, spaces, and subjects that readers tend to think of as available for bodily interaction—as haunting as that might be.

Is the Concert Hall Haunted?

The fin de siècle ghost stories of Falkner and Lee of course far surpass the limits of realism. The supernatural events they imagine, in which ghostly musicians haunt humans from beyond the grave, are inconceivable in the "real" world. Or are they? Isn't this what all music does, in a way? Doesn't all music, to some extent, make bodies quiver, blur senses of space and time, and draw listeners into embodied community with other listeners, players, and instruments and with the music itself? Are the sensations that John and Magnus experience really that different from what happens to everyone in a concert hall? When an audience member attends a performance of, say, Mahler's Symphony no. 2, they not only hear but physically feel the arrangements of notes that Mahler painstakingly teased out between 1888 and 1894, the bow strokes and woodwind tones from instruments crafted in Italy or Germany centuries ago, the fingerings and dynamics that performers have developed over the last century (and that have become entrenched as standard interpretations of the work), and the music emitted on stage at a spatial (and thus temporal) distance from them in their seats. Isn't every musical performance a kind of "haunting"?

It is perhaps in the unlikely realm of late-Victorian ghost stories, then, that readers can locate some of the most recognizable erotic experiences. While music does not make everyone "dissolve" with the intensity that Magnus does, it does usually reawaken sounds and sensations from other times and places and forge erotic connections that are invisible yet intensely palpable. As Le Guin writes, music always enables "reciprocal relationship[s] with someone no longer living."[93] In a world that often seems devoid of possibilities for intimacy and eroticism—or willfully committed to denying them—it is worth thinking of sound waves and musical vibrations as mechanisms for material contact with other kinds of bodies, other times, and other spaces. There might be hope for those whose pleasures are not as readily available in the "real" world; perhaps all one has to do is "suck in the sound."

Coda

Re-vitalizing Contemporary Classical Music

On April 16, 2022, the cast of *Saturday Night Live* (*SNL*) performed a skit called "Orchestra."[1] As the segment opens, viewers see a white male conductor (Mikey Day) in tails, scolding a collection of classical musicians, all dressed in black. "Stop, stop, stop," the conductor shakes his head. "This is a disaster. How are we going to perform Beethoven's Ninth Symphony in three hours without our *flautist*?" Enter Beverly Gags (Lizzo)—the "greatest flutist of all time, lips and lungs sent straight from God herself," according to her manager (Chris Redd). The conductor, though skeptical, allows Beverly—dressed in a bright-yellow blouse and sparkly black pants—to audition. She plays a few notes, all while twerking. While the conductor deems Beverly "clearly talented," he explains that she is not "quite our style"—a comment to which the wavy-haired, bespectacled harpist (Kate McKinnon) responds, frantically strumming all the while, "Don't be racist!" The conductor reminds everyone that "this is a traditional symphony, you know, one where you sit down and play" and asks Beverly to perform sitting down. Beverly concedes, but plays badly (shrill, out of tune, with incorrect notes) while seated. She explains, "The only way I can play the flute is if I'm twerking," before launching into an impassioned story about her musical awakening upon hearing the music of an ice cream truck as a child. "That beat had my body moving in ways that made me feel powerful," Beverly reminisces, "like I could do anything, even play the flute, and the rest is history." The conductor maintains that it would be "too distracting if one of us is twerking," to which the violinist (Aidy Bryant) triumphantly replies, "What if we *all* are twerking? We are the DeVry Institute Orchestra. We make music together. And if that means I gotta pop this booty for Beethoven, then so be it." The skit closes as all of the musicians, including the conductor, proceed to reprise the Beethoven theme while twerking.

The skit pokes fun at several elements of the Western classical music world—its traditionalism (the tuxedo tails and insistence on "sitting down"), its snobbery (the esteemed ensemble that hails from a famously predatory for-profit college), its pretensions (the affected, quasi-accented speech patterns of the white musicians and the insistence on the term *flautist* over *flutist*), and its racism (the rejection of twerking, of course, being a rejection of a Black cultural practice). As Garrett McQueen writes, the "Orchestra" skit highlights the ways in which "the perspectives and lived experiences of Black people and people of color, despite those individuals' ability to gain entry into predominantly white arts spaces, are continually marginalized."[2] More broadly, the skit also sheds light on the insistent erasure of the body—particularly the non-white, non-male body—from orchestral spaces. At the core of the racist, sexist, and sex-phobic treatment of Beverly, after all, is the Western classical music world's refusal to acknowledge that orchestras are made up of living, breathing bodies that sometimes like to move with the music and indeed play better while doing so. As ridiculous and perhaps problematic as the scene of collective twerking is (the white musicians' awkward movements could certainly be read as cultural appropriation or even mockery), the skit ultimately showcases the absurdity of pretending that there are no bodies onstage, that there is no "groove" to classical music, that the flute can't make one feel "powerful," that one can't "pop that booty" to Beethoven.

The body is a problem in the Western classical music world. Despite the centuries-old acoustical discoveries discussed in this book, as well as more recent neuroscientific research that traces what happens to our brains and bodies "on music," Western classical music culture persistently attempts to erase corporeal life.[3] This erasure largely began in the late nineteenth and early twentieth centuries—oddly, *after* acoustical scientists discovered the impact of music on human muscles and nerves. While the eighteenth-century concerts of, say, Mozart were lively affairs filled with food, drink, gossip, and revelry, toward the end of the nineteenth century, conductors and orchestra managers began to insist on increasingly rigid codes of conduct to regulate the bodies of players and listeners.[4] As a burgeoning population of middle-class listeners sought to perform their cultural capital, and composers like Wagner and Mahler began to idealize the "cult of the Work" above all else, the orchestral world witnessed an increased emphasis on what many now refer to as "concert etiquette."[5] For instance, E. M. Trevenen Dawson's guide "How to Behave at Concerts," published in the *Monthly Musical*

Record in 1898, instructed audience members at "philharmonic concert[s]" to maintain "a stolid, blank, or slightly *blasé* expression of countenance" and noted that "clapping should be decorous and staid."[6] This "etiquette" was so important, many critics believed, because it allowed performers and audiences to focus on the sanctity of music as a transcendent, spiritual entity. In 1897, the Chicago periodical the *Dial* praised the Chicago Orchestra conductor Theodore Thomas for his insistence on audience comportment; according to the *Dial*, conductors must tell the "average Philistine," "This masterpiece deserves your attention . . . for it has the power to raise you to a higher spiritual level."[7] The body must be ignored so that music's higher, spiritual powers can come to the foreground.

The emphasis on concert decorum in service of music's "higher spiritual level" persists today. Every few months, an incident of audience misbehavior—and, sometimes, a scolding by a player or conductor onstage—will make the news. The German bass-baritone Thomas Quasthoff once warned his audience, "Do not cough until the concert is ended. Because I love this music so much."[8] Similarly, in a 2013 review of a performance of the Polish composer Henryk Górecki's Symphony of Sorrowful Songs at the BBC Proms, *The Guardian*'s Martin Kettle wrote, "This was a rapt occasion. Górecki's solemn spell was only violated by the insistence of part of the Proms audience on applauding at the end of each movement."[9]

The rules governing audience behavior—all in service of preserving the music's "solemn spell"—are extensive. Concert attendees must arrive on time (or else be denied entry into the hall), remain still and silent, and refrain from clapping between movements.[10] Carnegie Hall provides complimentary throat lozenges in order to discourage audience coughing.[11] Former *Washington Post* classical music critic Anne Midgette's tongue-in-cheek "Ten Commandments of Concertgoing" hints at the religious fervor with which these codes are maintained: "Thou shalt not talk. Thou shalt not clap in the wrong places. Thou shalt not unwrap cough drops in crinkly paper during the music. Thou shalt not use thy cellphone."[12] Symphony orchestras often publish guides to concert etiquette. As Robin Bickerstaff Glover wrote for the Maryland Symphony Orchestra's "Symphony Etiquette" web page, "Don't talk, whisper, sing, hum, or move personal belongings. . . . Don't enter or exit the hall while a performance is in progress. . . . If you must leave your seat, do so quickly and quietly. . . . It is, however, appropriate to excuse yourself if you experience a prolonged bout of coughing or sneezing."[13] In 2014, the music critic Fiona Maddocks detailed a peculiarly torturous concertgoing experience: "I've trained myself to control most bodily urges (true!)

for the duration [of a symphony concert]."[14] But, during one performance by the Brodsky Quartet, she recalled, "I felt that terrible tightening of the muscles in the diaphragm as it pressed against my lungs. Any second now, I knew the glottis would open explosively. . . . I sucked and swallowed and stuffed a scarf into my mouth and wiped my tears. But still a strangulated explosion occurred. I spent the rest of the concert fearing another outburst, willing myself, reciting mantras."[15] It is no wonder, then, that in 1993, the League of American Orchestras identified the presence of "performance anxiety" among audience members as well as musicians.[16]

Just as listeners must suppress their coughs and constrict their bladders to preserve the music's "spell," performers must pretend they don't have bodies at all. From Wagner's famous "sunken orchestra" and "Bayreuth hush," to the standard "concert black" dress code in most symphony orchestras (tuxedos for men, all-black outfits for women), many of the long-held standards of classical music performance practice work to hide or homogenize the bodies onstage.[17] The practice of "blind" or "anonymous" auditions, in which prospective orchestra members play from behind a screen, exists to, as one March 2020 blog headline read, "Mitigat[e] the Body's Effect on Orchestral Jobs."[18] Cheng writes, "Anonymous orchestra auditions go to such comical lengths to bring bodies under erasure"; players must walk on carpets rather than hard floors to prevent their heels from clicking and refrain from breathing too deeply so as not to reveal their gender.[19] Similarly, describing the Boston Symphony Orchestra's anonymous audition process, the journalist Geoff Edgers notes, "Everything . . . is driven by a puritanical fervor for a single, guiding principle. All that matters is the music."[20]

Of course, it is certainly understandable that musicians might wish for attentive audiences, that audiences might yearn for focused listening experiences, and that everyone in the hall might desire moments of escape or transcendence from an often-untenable world. When a violinist agonizes for months over a particularly arduous double-stop, for instance, or a clarinetist achieves a smooth over-the-break interval leap, or a wind section carries out a delicately tuned ensemble passage, they want listeners to be able to hear it, free of the rustle of a program or a ringing cell phone. As musician Sasha Valeri Millwood writes, "Classical music is a fragile listening experience: most of the widely used instruments are not electronically amplified; and most of the repertoire . . . relies in part on the effort of quiet volumes and silence, an effect that is severely inhibited by extraneous noise and distraction. The formal concert setting ought to be a place where both audiences and performers can access the requisite focus to experience what the music has to offer."[21] Indeed, the rhetoric of music as an "escape" from the "real" world persists in discussions of music across genres and audiences. A brief glance at the e-commerce site Etsy reveals the enduring currency (metaphorical and literal) of the notion of music as a way to transcend an unbearable world. The retailer TreasuresDelightsEtc, for instance, sells a journal with the words "music is my escape" stenciled on the cover; as of September 2023, the item had been purchased and reviewed by over 6,030 customers.[22]

Yet such fantasies of musical escapism—and "assumptions about the intangibility and ephemerality of sound," as Cheng writes—are not apolitical.[23] There are profound consequences to erasing the body from the musical world in service of musical transcendence—especially when certain *kinds* of bodies are excluded or erased more than others. For performers, classical music's inattention to the body has actually meant that, for the most part, only one type of body—white, male, abled—has been able to achieve a sustained presence on the classical stage. It is difficult to imagine a more homogeneous artistic field in contemporary Western culture than classical music. As *New York Times* classical music critic Anthony Tommasini noted in July 2020, only one out of the 106 full-time players in the New York Philharmonic—the clarinetist Anthony McGill—is Black.[24] A 2014 study revealed that only 1.8 percent of full-time players in major US orchestras were Black, and 2.5 percent were Latino.[25] While the anonymous audition was initially set up to reduce discrimination and increase equity, many scholars believe that it has actually limited opportunities for deliberate and sustained antiracist hiring practices, serving as a convenient nod to "diversity" while actually propping up fantasies of "race-blind" meritocracy.[26] Cheng writes, "Accused of discriminating against black violinists or female percussionists? Just throw up a screen and purport to listen solely for merit by cutting all human variables out of the equation. By nobly claiming to judge people on musical ability alone, by dehumanizing them for all intents and purposes, we defer obligations to talk about trickier matters of humanity and prejudice."[27]

This pressure to "mitigate the body's effects" also means that performers must pretend their craft is *not* fundamentally an art of the body. Performers who attempt to showcase their bodies on the stage—especially women, nonbinary, and queer musicians and musicians of color—often face scorn from audiences and music critics. The pianist Yuja Wang—known for her daring costumes that often involve brightly colored, tight, and short dresses and high heels—has been scolded for sporting "stripper-wear."[28] The violinist Amadéus Leopold, who has been known to strut across the stage, lay on the piano, and dance as he plays, often incites "gasp[s]" in his "silver-haired" audience members.[29]

The classical music world's erasure of the body can also result in devastating health consequences for the very bodies it erases. Classical musicians must often work through debilitating injuries, including repetitive strain injuries (RSIs) as a result of grueling hours of practice each day. According to Dr. Serap Bastepe-Gray of the Johns Hopkins Center for Music and Medicine, four in five professional musicians are injured over the course of their careers.[30] Yet performers—particularly women and people of color—must often go to great lengths to prove that they will not be limited by their bodies. Before she joined the Munich Philharmonic, the female trombonist Abbie Conant—"alone among her male peers"—was forced to undergo "degrading medical, physical, and musical examinations to prove her fitness."[31] The COVID-19 pandemic further underscored the bodily dangers of musical performance.[32] One of the earliest "superspreader" events occurred at a community choir rehearsal in Washington, and scientists have since released studies on how orchestral musicians should be positioned onstage to reduce the transmission of aerosols.[33] As of 2022, the legendary violinist Itzhak Perlman still requested that all audience members wear masks at his indoor performances.[34] If players can contract COVID-19 or other airborne illnesses onstage, then no one can deny that orchestral performance is an art of living, moving, *breathing* bodies.

For audience members, too, the privileging of "the music itself" is often accompanied by insidious, exclusionary ideas about manners, etiquette, comportment, and "good behavior"—deeply classist and racist ideologies that are based on, Kirsty Sedgman writes, centuries of imperialist "civilising discourse" in which certain groups "wor[k] collectively to assert themselves as the obvious arbiters of reasonable behaviour."[35] Attempts to curate the sublime concert hall experience often alienate those without the means or opportunities to gain insider knowledge about concert etiquette. As the music journalist Lukas Krohn-Grimberghe states,

> Most of us believe that classical music is a universal language—especially when it's instrumental—as it speaks to our hearts and invites us all to fill it with our own imaginations, meanings, hopes, and dreams. It can serve as a weighted blanket to soothe us, or as a refreshing drink that invigorates and brings joy. In an ideal world, classical music would be for everyone, irrespective of their social or economic status—or the color of their

skin. However, classical music is actually used (consciously and unconsciously) as a signifier for social class, and thus, also for racial segregation. . . . Classical music, thus, becomes a refuge for white elites—a space to breathe—a luxury black listeners and musicians often don't have.[36]

The veneration of "the music itself" and the insistence on the silent audience also excludes concertgoers with disabilities or illnesses that may render their bodies "disruptive" in a concert setting. What Hannah Simpson writes of the theatre is perhaps even more true of the classical concert hall:

> The cult of the quiet audience presents a sometimes insurmountable challenge to the neurodivergent spectator, whose cognitive and/or physical functioning may mean that she cannot guarantee that her body will remain quiet during the length of a performance. . . . Audience noise that signals only an alternatively functioning body is often condemned as equally inappropriate or disrespectful in the theatre auditorium: the verbal tic or motor convulsion of the person with Tourette's syndrome, the repetitive tapping of the individual with obsessive-compulsive disorder, or the self-comforting rocking of the autistic child.[37]

In the logic of traditional concert etiquette, noisy bodies threaten to disrupt the purity of the acoustical space and the elusive contact with the exalted musical work.

Recently, activists in classical music and theatre performance spaces have urged a vigorous interrogation of some of the fields' most deeply entrenched practices, encouraging organizations to reexamine their recruitment strategies, audience engagement efforts, and performance practices. They query: What if classical music took the body seriously as something that experiences and produces art, as something that adds to rather than distracts from the pleasures of the performance? What would happen if everyone stopped pretending that performers and listeners don't have muscles that twitch, fingers that get tired, diaphragms that contract, or bladders that must be emptied? Whom might the field be able to include? What kinds of engagements with music would this enable? What experiences could performers and audiences begin to imagine—and celebrate?

For audience members, this might mean more "accessible" or "relaxed" performances, such as those put on by Touretteshero or the Relaxed Performance Project in the UK.[38] It might mean more moments like that which took place at a 2019 performance by Boston's Handel and Haydn Society, when a child with autism joyfully exclaimed "Wow!" at the concert's conclusion.[39] A more embodied classical music world might involve opening up performances to first-time concertgoers, families with children, and neurodivergent, disabled, or sick people whose bodies might indeed shift the soundscape of the hall. Or it could involve a new kind of engagement with music for *everyone*, in which listeners are free to react as the music moves them, to share their pleasures, to communicate their excitement. Perhaps audiences might even forge new kinds of listening communities through shared affects—communities in which, as Simpson writes of the relaxed theatre audience, members "acknowledg[e] and welcom[e] this recognition of others in one's own world, in a dynamic, intersubjective process of mutual accommodation."[40]

A deeper attention to the body can usher in new possibilities for performers as well. A growing body of musicological research emphasizes the importance of corporeal attunement for classical singers and instrumentalists. A study at the Danish National Academy of Music, for instance, revealed that attending to the body can help performers develop more "dynamic and free" performances.[41] The study examined workshops that emphasized "embodiment of the music—physical awareness, gestures, facial expressions, and eyes," with "the aim being having the physicality support the interpretation and encouraging the playfulness of the performer."[42] In a 2012 masterclass at Yale, the violin/piano duo Igudesman and Joo taught students to harness the powers of their bodies to improve their performances. As attendee Astrid Baumgardner wrote,

> Igudesman and Joo have learned how to use their bodies effectively and have discovered the freedom that comes from movement. Their performances are laced with physical antics: Joo's playing the piano upside down while lying on the floor, Igudesman's playing the violin while kneeling and lying down or dancing around the stage. Their facial expressions add to reinforce their joy in making fun with the music. I love it when performers show me through their bodies how they feel about the music.[43]

Refusing to pretend that the performer is just a vehicle for "the music itself" could also mean including *more* kinds of bodies on the contemporary concert hall stage. Several musicians and musicologists propose that abandoning anonymous auditions, for instance, could help create orchestras that better reflect and serve their communities and erase meritocratic myths that systemically exclude women and people of color. As McGill and Afa Dworkin wrote in the *New York Times* in September 2020, orchestras need to develop a "transparent definition of what is needed of a musician to join and be tenured in a reimagined collective of artists."[44] McGill and Dworkin suggest, "Invest in expanding the process to assess teaching, community ambassadorship and their ability to act as a public-facing artist. Invest in the right experts who will understand race, culture and talent, and can competently guide your selection process."[45]

Acknowledging classical music as an embodied art form would also make more space for nonwhite, female, queer, and nonbinary performers like Wang and Leopold, whose performance practices are fundamentally based in their insistent refusals to erase their performing bodies. When asked about her lavish, risqué outfits, Wang told *The Guardian*, "If the music is beautiful and sensual, why not dress to fit?"[46] Leopold's unorthodox performances have helped others reimagine what a virtuoso can look and act like. As madison moore writes, "The moment Leopold steps out onstage, perhaps the only 'safe' space, audiences typically respond with an audible gasp, he noted, and for him that gasp circles the way fabulousness changes the energy in a room. That's when the real performance begins. . . . Leopold's art pinpoints in no uncertain terms that fabulousness is something embodied and queer—an aesthetic—one rooted in certain kinds of creative agency, where extravagant self-expression is a dangerous political gesture."[47] Celebrating the affordances of the body onstage can thus invite new possibilities and pleasures for both audience members and performers in a cultural sphere long defined by exclusion and regulation.

The notion that a musical encounter can be fundamentally enlivened when performers harness the powers and pleasures of their bodies, and when audience members feel free to lean into the raptures they experience while listening, would have been familiar to the Victorians discussed in this book. Nineteenth-century acoustical scientists and the writers they inspired showed that tuning into the musical body—the hands that hold the violin, the ears that receive sonic vibrations, the fingers that stroke the keys, the spine that tingles in response to a sound—allows both performers and listeners to express and experience a wide range of desires, pleasures, and sensations. For many Victorian writers, conceptualizing music not in a vacuum of aesthetic sublimity but as an enlivened and embodied art form enabled radical reimaginations of gender, sexuality, and intimacy.

Today's classical music world would benefit from opportunities to think along similar lines. By refusing to pretend that all that is at stake at a symphony concert is "the music itself," orchestra directors, boards, critics, and players might more readily acknowledge that ensembles are composed of people with bodies that do not all look, act, or feel the same way; bodies that might need to cough or shuffle or pee; bodies that ache after playing or ache *to* play; bodies that might want to express themselves or engage with performances in unexpected ways. It is in the materiality of performance that music's most radical potential lies.

Notes

Introduction

1. Amy Levy, "Sinfonia Eroica," in *A Minor Poet and Other Verses* (London: T. Fisher Unwin, 1884), 59.

2. Levy, 60–61.

3. Levy, 59–60.

4. Levy, 60.

5. Phyllis Weliver, introduction to *The Figure of Music in Nineteenth-Century British Poetry* (Aldershot: Ashgate, 2005), 1n1, 4; Linda K. Hughes, " 'Phantoms of Delight': Amy Levy and Romantic Men," in *Decadent Romanticism: 1780–1914*, ed. Kostas Boyiopoulos and Mark Sandy (Aldershot: Ashgate, 2015), 175; Fraser Riddell, *Music and the Queer Body in English Literature at the* Fin de Siècle (Cambridge: Cambridge University Press, 2022), 200–202. See also David Deutsch, *British Literature and Classical Music: Cultural Contexts, 1870–1945* (London: Bloomsbury, 2015), 146.

6. Eve Kosofsky Sedgwick, *Touching Feeling: Affect, Pedagogy, Performativity* (Durham, NC: Duke University Press, 2003), 150–51.

7. Levy, "Sinfonia," 59. Weliver and Hughes both acknowledge the poem's engagement with music as a physical force but focus on the music's connections to the speaker's private desires. Hughes, " 'Phantoms,' " 170; Weliver, introduction, 3–4.

8. For more on the English Musical Renaissance, see Phyllis Weliver, *The Musical Crowd in English Fiction: Class, Culture, and Nation* (London: Palgrave Macmillan, 2006), 3.

9. Hermann von Helmholtz, *On the Sensations of Tone as a Physiological Basis for the Theory of Music*, trans. Alexander Ellis, 3rd ed. (London: Longmans, Green, 1895), 3.

10. Hector Chomet, *The Influence of Music on Health and Life*, trans. Laura A. Flint (New York: G. P. Putnam's Sons, 1875), 192–93; "Physiological Effects of Music," *Canterbury Journal and Farmer's Gazette*, no. 3017 (May 26, 1894): 6.

11. John Tyndall, *Sound*, 1st ed. (London: Longmans, Green, 1867), 49.

12. Benjamin Steege, *Helmholtz and the Modern Listener* (Cambridge: Cambridge University Press, 2012), 24.

13. Shannon Draucker, "Hearing, Sensing, Feeling Sound: On Music and Physiology in Victorian England, 1857–1894," *BRANCH: Britain, Representation and Nineteenth-Century History*, edited by Dino Franco Felluga, extension of *Romanticism and Victorianism on the Net*, https://branchcollective.org/?ps_articles=shannon-draucker-hearing-sensing-feeling-sound-on-music-and-physiology-in-victorian-england-1857-1894.

14. Riddell, *Music*, 201.

15. Audre Lorde, "Uses of the Erotic: The Erotic as Power," in *Sister Outsider: Essays and Speeches* (Berkeley, CA: Crossing Press, 1984), 55–57; Tim Dean, *Beyond Sexuality* (Chicago: University of Chicago Press, 2000); Tim Dean, "The Biopolitics of Pleasure," *South Atlantic Quarterly* 111, no. 3 (2012): 477–96; Ela Przybylo, *Asexual Erotics: Intimate Readings of Compulsory Sexuality* (Columbus: Ohio State University Press, 2019). Recently, Joan Morgan has emphasized the need to retain the sexual dimensions of the erotic, particularly in Black feminist contexts: "I want an erotic that demands space be made for honest bodies that like to also *fuck*." Joan Morgan, "Why We Get Off: Moving Towards a Black Feminist Politics of Pleasure," in *Pleasure Activism: The Politics of Feeling Good*, ed. adrienne maree brown (Chico, CA: AK Press, 2019), 90.

16. Eve Kosofsky Sedgwick, *Epistemology of the Closet* (Berkeley: University of California Press, 1990), 8; Cathy Cohen, "Punks, Bulldaggers, and Welfare Queens: The Radical Potential of Queer Politics?" *GLQ* 3, no. 4 (May 1997): 437–65; Jack Halberstam, *In a Queer Time and Place: Transgender Bodies, Subcultural Bodies* (New York: New York University Press, 2005), 6; Elizabeth Freeman, *Time Binds: Queer Temporalities, Queer Histories* (Durham, NC: Duke University Press, 2010); Sara Ahmed, *Queer Phenomenology: Orientations, Objects, Others* (Durham, NC: Duke University Press, 2006).

17. Levy, "Sinfonia," 60.

18. Levy, 61.

19. Levy, 60.

20. Suzanne Cusick, "On a Lesbian Relationship with Music: A Serious Effort Not to Think Straight," in *Queering the Pitch*, ed. Philip Brett and Elizabeth Wood (New York: Routledge, 1994), 78–9, emphasis in original.

21. Levy, "Sinfonia," 60–61, emphasis mine.

22. Levy, 60–61.

23. Sara Ahmed, *What's the Use?* (Durham, NC: Duke University Press, 2019), 199.

24. For other discussions of (and calls for) feminist methodologies in Victorianist criticism, see Alison Booth, "Feminism," *Victorian Literature and Culture* 46, no. 3/4 (Fall/Winter 2018): 691–97; Jill Ehnenn, "From 'We Other Victorians' to

'Pussy Grabs Back': Thinking Gender, Thinking Sex, and Feminist Methodological Futures in Victorian Studies," *Victorian Literature and Culture* 41, no. 1 (2019): 35–62; Talia Schaffer, "Feminism and the Canon," in *The Routledge Companion to Victorian Literature*, ed. Dennis Denisoff and Talia Schaffer (New York: Routledge, 2020), 273–83.

25. Jodi Taylor, "Taking It in the Ear: On Music-Sexual Synergies and the (Queer) Possibility That Music Is Sex," *Continuum* 26, no. 4 (August 2012): 612.

26. Here I refer to what Julian Johnson calls the "popular use" definition of the term "classical music" to denote "music from a wider historical period (say, from around 1600 to the present day) now associated with performance in a concert hall or opera house." Julian Johnson, *Who Needs Classical Music? Cultural Choice and Musical Value* (Oxford: Oxford University Press, 2002), 6. I also deliberately use the term "Western classical music" to resist the implicit, erroneous, and often-unnamed associations of the term "classical music" with only Western European (and mostly white) cultures, despite centuries-old traditions of, for example, Chinese or Indian classical music. For more on the term "Western classical music," see Philip Ewell, *On Music Theory, and Making Music More Welcoming for Everyone* (Ann Arbor: University of Michigan Press, 2023); Garrett McQueen, "The Power (and Complicity) of Classical Music," *Your Classical* (June 5, 2020), https://www.yourclassical.org/story/2020/06/05/the-power-and-complicity-of-classical-music.

27. Ronjaunee Chatterjee, Alicia Mireles Christoff, and Amy R. Wong, "Introduction: Undisciplining Victorian Studies," *Victorian Studies* 62, no. 3 (Spring 2020): 379.

28. Chatterjee, Christoff, and Wong, "Introduction," 378.

29. For more, see Jeffrey Green, *Samuel Coleridge-Taylor, a Musical Life* (London, Routledge, 2011); Nina Sun Eidsheim, *The Race of Sound: Listening, Timbre, & Vocality in African American Music* (Durham, NC: Duke University Press, 2019).

30. Kadji Amin, *Disturbing Attachments: Genet, Modern Pederasty, and Queer History* (Durham, NC: Duke University Press, 2017), 11.

31. Amin, *Disturbing Attachments*, 8.

32. Kristin Mahoney, *Queer Kinship After Wilde: Transnational Decadence and the Family* (Cambridge: Cambridge University Press, 2022), 11–12.

33. Sedgwick, *Touching*, 150–51.

34. José Esteban Muñoz, *Cruising Utopia: The Then and There of Queer Futurity* (New York: New York University Press, 2009), 1.

35. Lorde, "Uses of the Erotic."

36. John Picker, *Victorian Soundscapes* (Oxford: Oxford University Press, 2003), 7.

37. Gillian Beer, *Open Fields: Science in Cultural Encounter* (Oxford: Oxford University Press, 1996), 245, 255.

38. Picker, *Soundscapes*, 92. See also Jay Clayton, "Hacking the Nineteenth Century," in *Victorian Afterlife: Postmodern Culture Rewrites the Nineteenth Century*,

ed. John Kucich and Dianne F. Sadoff (Minneapolis: University of Minnesota Press, 2000), 186–210; David Sweeney Coombs, *Reading with the Senses in Victorian Literature and Science* (Charlottesville: University of Virginia Press, 2019), ch. 5.

39. Helmholtz, *Sensations*, 3–4, emphasis in original.

40. Helmholtz, 4.

41. William Cohen, *Embodied: Victorian Literature and the Senses* (Minneapolis: University of Minnesota Press, 2008); Nicholas Dames, *The Physiology of the Novel: Reading, Neural Science, and the Form of Victorian Fiction* (Oxford: Oxford University Press, 2007); Benjamin Morgan, *The Outward Mind: Materialist Aesthetics in Victorian Science and Literature* (Chicago: University of Chicago Press, 2017); Jason Rudy, *Electric Meters: Victorian Physiological Poetics* (Athens: Ohio University Press, 2009).

42. Lorraine Daston and Peter Galison, *Objectivity* (Brooklyn: Zone Books, 2007).

43. Michel Foucault, *The History of Sexuality*, vol. 1, trans. Robert Hurley (New York: Vintage Books, 1990), 42, 44.

44. Carolyn Burdett, "Sexual Selection, Automata and Ethics in George Eliot's *The Mill on the Floss* and Olive Schreiner's *Undine* and *From Man to Man*," *Journal of Victorian Culture* 14, no. 1 (January 2009): 29; Dustin Friedman, *Before Queer Theory: Victorian Aestheticism and the Self* (Baltimore: Johns Hopkins University Press, 2019), 14.

45. Riddell, *Music*, 48.

46. Angelique Richardson, *Love and Eugenics in the Late Nineteenth Century: Rational Reproduction and the New Woman* (Oxford: Oxford University Press, 2003). For discussions of the racial politics of Victorian evolutionary, physiological, and acoustical science, see Erica Fretwell, *Sensory Experiments: Psychophysics, Race, and the Aesthetics of Feeling* (Durham, NC: Duke University Press, 2020); Alexandra Hui, *The Psychophysical Ear: Musical Experiments, Experimental Sounds, 1840–1910* (Cambridge, MA: MIT Press, 2013); Bennett Zon, *Evolution and Victorian Musical Culture* (Cambridge: Cambridge University Press, 2017); Kyla Schuller, *The Biopolitics of Feeling: Race, Sex, and Science in the Nineteenth Century* (Durham, NC: Duke University Press, 2018); William Cheng, *Loving Music Till It Hurts* (Oxford: Oxford University Press, 2019). I further discuss the racism of nineteenth-century acoustical science in chapter 1.

47. Ahmed, *What's the Use?*, 197.

48. Elizabeth A. Wilson, *Psychosomatic: Feminism and the Neurological Body* (Durham, NC: Duke University Press, 2004), 13.

49. Elizabeth Grosz, *Volatile Bodies: Toward a Corporeal Feminism* (Bloomington: Indiana University Press, 1994), ix; Jay Prosser, *Second Skins: The Body Narratives of Transsexuality* (New York: Columbia University Press, 1998), 62; Angela Willey, "Biopossibility: A Queer Feminist Materialist Science Studies Manifesto, with Special Reference to the Question of Monogamous Behavior," in *Feminist and Queer Theory: An Intersectional and Transnational Reader*, ed. L. Saraswati and Barbara

Shaw (Oxford: Oxford University Press, 2021), 510–11; Angela Willey, *Undoing Monogamy: The Politics of Science and the Possibilities of Biology* (Durham, NC: Duke University Press, 2016). See also Susan Stryker, "Transgender Studies: Queer Theory's Evil Twin," in Saraswati and Shaw, *Feminist and Queer Theory*, 70–72; Sara Ahmed, "Open Forum: Imaginary Prohibitions; Some Preliminary Remarks on the Founding Gestures of the 'New Materialism,'" *European Journal of Women's Studies* 15, no. 1 (2008): 23–39; Noela Davis, "New Materialism and Feminism's Anti-biologism: A Response to Sara Ahmed," *European Journal of Women's Studies* 16, no. 1 (2009): 67–80; Deboleena Roy, "Somatic Matters: Becoming Molecular in Molecular Biology," *Rhizomes*, no. 14 (Summer 2007), http://www.rhizomes.net/issue14/roy/roy.html.

50. Anna Peak, "The Condition of Music in Victorian Scholarship," *Victorian Literature and Culture* 44, no. 2 (2016): 423–37; Phyllis Weliver, "A Score of Change: Twenty Years of Critical Musicology and Victorian Literature," *Literature Compass* 8, no. 10 (2011): 776–94.

51. Nina Sun Eidsheim, *Sensing Sound: Singing and Listening as Vibrational Practice* (Durham, NC: Duke University Press, 2015), 2.

52. Claire Jarvis, *Exquisite Masochism: Marriage, Sex, and the Novel Form* (Baltimore: Johns Hopkins University Press, 2016), vii, 18; Doreen Thierauf, "The Hidden Abortion Plot in George Eliot's *Middlemarch*," *Victorian Studies* 56, no. 3 (Spring 2014): 479–89; Doreen Thierauf, "*Daniel Deronda*, Marital Rape, and the End of Reproduction," *Victorian Review* 43, no. 2 (Fall 2017): 247–69; Livia Arndal Woods, "Now You See It: Concealing and Revealing Pregnant Bodies in *Wuthering Heights* and *The Clever Woman of the Family*," *Victorian Network* 6, no. 1 (Summer 2015): 32–54.

53. William Cohen, *Sex Scandal: The Private Parts of Victorian Fiction* (Durham, NC: Duke University Press, 1996), 3, emphasis in original.

54. Gemma Moss, "Classical Music and Literature," in *Sound and Literature*, ed. Anna Snaith (Cambridge: Cambridge University Press, 2020), 92. See also Emma Sutton, "'The Music Spoke for Us': Music and Sexuality in *Fin-de-siècle* Poetry," in Weliver, *Figure*, 214–15; Joe Law, "The 'Perniciously Homosexual Art': Music and Homoerotic Desire in *The Picture of Dorian Gray* and Other *Fin-de-Siècle* Fiction," in *The Idea of Music in Victorian Fiction*, ed. Sophie Fuller and Nicky Losseff (Aldershot: Ashgate, 2004), 173–96.

55. Law, "'Perniciously,'" 196, 177.

56. Oscar Wilde, *The Picture of Dorian Gray* (Oxford: Oxford World's Classics, 2008), 140; Law, "'Perniciously,'" 178.

57. Oscar Wilde, *De Profundis* (London: Methuen, 1912).

58. Elisabeth Le Guin, *Boccherini's Body: An Essay in Carnal Musicology* (Berkeley: University of California Press, 2006), 5; Suzanne Cusick, "On Musical Performances of Sex and Gender," in *Audible Traces: Gender, Identity, and Music*, ed. Elaine Barkin and Lydia Hamessley (Zurich: Carciofoli Verlagshaus, 1999), 42–43;

Dana Baitz, "Toward a Trans* Method in Musicology," in *The Oxford Handbook of Music and Queerness*, ed. Fred Everett Maus and Sheila Whiteley (Oxford: Oxford University Press, 2022), 370–72. See also Suzanne Cusick, "Feminist Theory, Music Theory, and the Mind/Body Problem," *Perspectives of New Music* 32, no. 1 (Winter 1994): 25; Philip Auslander, "Musical Personae," *TDR: The Drama Review* 50, no. 1 (2006): 100–119; Carolyn Abbate, "Music—Drastic or Gnostic?" *Critical Inquiry* 30, no. 3 (Spring 2004): 505–36; Nicholas Cook, *Beyond the Score: Music as Performance* (Oxford: Oxford University Press, 2013); Youn Kim and Sander L. Gilman, "Contextualizing Music and the Body: An Introduction," in *The Oxford Handbook of Music and the Body*, ed. Youn Kim and Sander L. Gilman (Oxford: Oxford University Press, 2019), 1–20.

59. Sophie Fuller, "Dead White Men in Wigs," in *Girls! Girls! Girls! Essays on Women and Music*, ed. Sarah Cooper (New York: New York University Press, 1996), 22.

60. Ruth Solie, "No 'Land without Music' After All," *Victorian Literature and Culture* 32, no. 1 (2004): 261.

61. Kirsty Sedgman, *The Reasonable Audience: Theatre Etiquette, Behaviour Policing, and the Live Performance Experience* (London: Palgrave Macmillan, 2018); James Johnson, *Listening in Paris: A Cultural History* (Berkeley: University of California Press, 1995); Charles Edward McGuire, *Music and Victorian Philanthropy: The Tonic Sol-Fa Movement* (Cambridge: Cambridge University Press, 2009); Mary Burgan, "Heroines at the Piano: Women and Music in Nineteenth-Century Fiction," *Victorian Studies* 30, no. 1 (Autumn 1986): 51–76; Derek B. Scott, *The Singing Bourgeois: Songs of the Victorian Drawing Room and Parlour* (Aldershot: Ashgate, 2001).

62. Burgan, "Heroines," 51.

63. Phyllis Weliver, *Women Musicians in Victorian Fiction, 1860–1900: Representations of Music, Science and Gender in the Leisured Home* (Aldershot: Ashgate, 2000), 1.

64. Alisa Clapp-Itnyre, *Angelic Airs, Subversive Songs: Music as Social Discourse in the Victorian Novel* (Athens: Ohio University Press, 2002), xvii.

65. Riddell, *Music*, 48, 98.

66. Lorde, "Uses of the Erotic," 54–57.

67. Przybylo, *Asexual*, 20.

68. Ardel Haefele-Thomas, "Introduction: Trans Victorians," *Victorian Review* 44, no. 1 (Spring 2018): 33; Lisa Hager, "A Case for a Trans Studies Turn in Victorian Studies: 'Female Husbands' of the Nineteenth Century," *Victorian Review* 44, no. 1 (Spring 2018): 37–54; Simon Joyce, *LGBT Victorians: Sexuality and Gender in the Nineteenth-Century Archives* (Oxford: Oxford University Press, 2022).

69. For work on queerness and antinormativity (or "anti-antinormativity"), see Robyn Wiegman and Elizabeth A. Wilson, "Introduction: Antinormativity's Queer Conventions," *differences* 26, no. 1 (2015): 1–25; Jasbir Puar, *Terrorist Assemblages: Homonationalism in Queer Times* (Durham, NC: Duke University Press, 2007); Jean Bessette, "Queer Rhetoric in Situ," *Rhetoric Review* 35, no. 2 (2016): 148–64; Sara

Ahmed, *The Cultural Politics of Emotion* (London: Routledge, 2004); Heather Love, "Doing Being Deviant: Deviance Studies, Description, and the Queer Ordinary," *differences* 26, no. 1 (2015): 74–95.

70. Deborah Lutz, *Pleasure Bound: Victorian Sex Rebels and the New Eroticism* (New York: Norton, 2011); Ellen Bayuk Rosenman, *Unauthorized Pleasures* (Ithaca, NY: Cornell University Press, 2003); Sharon Marcus, *Between Women: Friendship, Desire, and Marriage in Victorian England* (Princeton, NJ: Princeton University Press, 2007); Holly Furneaux, *Queer Dickens: Erotics, Families, Masculinities* (Oxford: Oxford University Press, 2013); Abigail Joseph, *Exquisite Materials: Episodes in the Queer History of Victorian Style* (Newark: University of Delaware Press, 2019); Talia Schaffer, *Romance's Rival: Familiar Marriage in Victorian Fiction* (Oxford: Oxford University Press, 2016); Friedman, *Before Queer Theory*; Mahoney, *Queer Kinship*.

71. Mary Augusta Ward, *Robert Elsmere* (Brighton, UK: Victorian Secrets, 2013), 253; M. E. Francis, *The Duenna of a Genius* (Leipzig: Bernhard Tauchnitz, 1899), 114.

72. Jodi Taylor, "Sound Desires: Auralism, the Sexual Fetishization of Music," in Maus and Whiteley, *Oxford Handbook of Music and Queerness*, 277–94; Judith Peraino, "The Same, but Different: Sexuality and Musicology, Then and Now," *Journal of the American Musicological Society* 66, no. 3 (Fall 2013): 825–30; Fred Everett Maus, "'What If Music IS Sex?': Suzanne Cusick and Collaboration," *Radical Musicology* 7 (2019), http://www.radical-musicology.org.uk/2019/Maus.htm.

73. Suzanne Cusick, "Response: 'This Song Is for You,'" *Journal of the American Musicological Society* 66, no. 3 (Fall 2013): 862.

74. Paula Gillett, *Musical Women in England, 1870–1914: "Encroaching on All Man's Privileges"* (New York: St. Martin's, 2000), 98, 140.

75. Erin Spampinato, "Rereading Rape in the Critical Canon: Adjudicative Criticism and the Capacious Conception of Rape," *differences* 32, no. 2 (2021): 142.

76. This is the phrase used in the 1533 Buggery Act, which remained in place in one form or another throughout the nineteenth century. Amanda Mordavsky Caleb, introduction to *Teleny; or, the Reverse of the Medal*, ed. Amanda Mordavsky Caleb (Richmond, VA: Valancourt Books, 2010). The 1861 Offences Against the Person Act contained a section on "Unnatural Offences" that described "buggery" as an "abominable crime." Offences Against the Person Act, 1861, in *Nineteenth-Century Writings on Homosexuality: A Sourcebook*, ed. Chris White (New York: Routledge, 1999), 44.

77. Freeman, *Time Binds*, 95.

Chapter One

1. "A Holiday Afternoon with Professor Tyndall," *Daily News*, no. 8635 (December 29, 1873): 5.

2. "Holiday," 5.

3. "Holiday," 5.
4. "Holiday," 5.
5. Tyndall, *Sound*, 1st ed., 49; Helmholtz, *Sensations*, 7.
6. Helmholtz, *Sensations*, 3.
7. Tyndall, *Sound*, 1st ed., 158; Myles W. Jackson, *Harmonious Triads: Physicists, Musicians, and Instrument Makers in Nineteenth-Century Germany* (Cambridge, MA: MIT Press, 2006), 13–44.
8. Tyndall, *Sound*, 1st ed., 135.
9. Leigh Eric Schmidt, *Hearing Things: Religion, Illusion and the American Enlightenment* (Cambridge, MA: Harvard University Press, 2000), 13.
10. Edward J. Gillin, *Sound Authorities: Scientific and Musical Knowledge in Nineteenth-Century Britain* (Chicago: University of Chicago Press, 2021), 4.
11. Gillin, *Sound Authorities*, 25–82; John Herschel, "Sound," in *Encyclopedia Metropolitana* (London: William Clowes and Sons, 1845), 810.
12. For more on Helmholtz's studies of optics, see Steege, *Helmholtz*; David Cahan, *Helmholtz: A Life in Science* (Chicago: University of Chicago Press, 2018). See also Stephan Vogel, "Sensations of Tone, Perception of Sound, and Empiricism: Helmholtz's Physiological Aesthetics," in *Hermann von Helmholtz and the Foundations of Nineteenth-Century Science*, ed. David Cahan (Berkeley: University of California Press, 1993), 259–90.
13. Cahan, *Life*, 196–97.
14. Gillin, *Sound Authorities*, 12; Journal of John Tyndall, October 1858–July 1871, RI MS JT/2/10, Royal Institution of Great Britain Archives, Royal Institution of Great Britain, London (hereafter referred to as Royal Institution).
15. Qtd. in Cahan, *Life*, 262.
16. Steege, *Helmholtz*, 43, 195; Hermann von Helmholtz, syllabus, "A Course of Six Lectures on the Natural Law of Conservation of Energy by Professor H. Helmholtz, F.R.S," RI MS GB 03/104-6, Royal Institution.
17. Helmholtz, *Sensations*, 7.
18. Helmholtz, 7–8.
19. Tyndall, *Sound*, 1st ed., 50–51. See also Gillin, *Sound Authorities*, 5.
20. Helmholtz, *Sensations*, 36.
21. Helmholtz, 10, 36–39.
22. Helmholtz, 10, 36.
23. Helmholtz, *Sensations*, 39, 48–49.
24. Steege, *Helmholtz*, 186.
25. David Cahan, "Helmholtz and the British Scientific Elite: From Force Conversation to Energy Conservation," *Royal Society Journal of the History of Science* 66 (2012): 55.
26. Tyndall, Journal.
27. *Notices of the Proceedings at the Meetings of the Members of the Royal Institution of Great Britain*, vols. 7–9 (London: William Clowes and Sons, 1875–1881); Tyndall, *Sound*, 3rd ed. (New York: D. Appleton, 1898), 402.

28. Tyndall, *Sound*, 1st ed., 4.
29. Tyndall, *Sound*, 1st ed., 4–5.
30. Roger Matthew Grant, "Peculiar Attunements: Comic Opera and Enlightenment Mimesis," *Critical Inquiry* 43, no. 2 (2017): 563; Shelley Trower, *Senses of Vibration: A History of the Pleasure and Pain of Sound* (London: Continuum, 2012), 13; Jerome McGann, *The Poetics of Sensibility* (Oxford: Oxford University Press, 1996); Susan Bernstein, "On Music Framed: The Eolian Harp in Romantic Writing," in Weliver, *Figure*, 70–84.
31. According to Trower, Hartley did not move beyond a figurative link between strings and nerves, believing it "highly absurd" that the nerves should actually vibrate like musical strings. Trower, *Vibration*, 9, 17–18. See also Roger Matthew Grant, *Peculiar Attunements: How Affect Theory Turned Musical* (New York: Fordham University Press, 2020): 78–79, 92; Wayne D. Bowman, *Philosophical Perspectives on Music* (Oxford: Oxford University Press, 1998), 74.
32. Edmund Burke, *A Philosophical Enquiry into the Origin of Our Ideas of the Sublime and Beautiful*, ed. James T. Boulton (South Bend, IN: University of Notre Dame Press, 1986), 122–23.
33. Herbert Spencer, "The Origin and Function of Music," in *Essays: Scientific, Political, and Speculative*, vol. 2 (London: Williams and Norgate, 1891), 404, 411.
34. Spencer, "Origin," 404.
35. See Peter Kivy, *Music, Language, and Cognition: And Other Essays in the Aesthetics of Music* (Oxford: Oxford University Press, 2007), 18–19; Bennett Zon, *Music and Metaphor in Nineteenth-Century British Musicology* (Aldershot: Ashgate, 2000).
36. Kivy, *Music*, 17.
37. Gillin, *Sound Authorities*, 12.
38. William Robert Wilde, *Practical Observations on Aural Surgery and the Nature and Treatment of Diseases of the Ear* (Philadelphia: Blanchard and Lea, 1853), 31.
39. Jonathan Sterne, *The Audible Past: Cultural Origins of Sound Reproduction* (Durham, NC: Duke University Press, 2003), 54–55.
40. Jennifer Esmail, *Reading Victorian Deafness: Signs and Sounds in Victorian Literature and Culture* (Athens: Ohio University Press, 2013), 165.
41. Wilde, *Practical Observations*, 30–1.
42. John Harrison Curtis, *A Treatise on the Physiology and Diseases of the Ear* (London: John Anderson, 1819), 7, 25.
43. Curtis, 7, 25.
44. Curtis, 1–15; John Harrison Curtis, *On the Cephaloscope and Its Uses in the Discrimination of the Normal and Abnormal Sounds in the Organ of Hearing* (London: John Churchill, 1842), 48–9.
45. Curtis, *Treatise*, 26.
46. Helen Keller, *The Story of My Life* (New York: Doubleday, Page, 1905), 6, 31.
47. Helmholtz, *Sensations*, 4–5.

48. Helmholtz, *Sensations*, vii. Steege notes that Helmholtz departed from many of his contemporary German philosophers, who advocated for a *Geist* (or "mind") "as the bearer of intellectual and spiritual autonomy." Steege, *Helmholtz*, 58. Steege suggests that though Helmholtz was "neither fully idealist . . . nor fully materialist-sensualist," his "discussion of the 'material ear' far outweighs in words that of the 'mental' or 'spiritual' ear." Steege, *Helmholtz*, 129, 73. See also Hermann von Helmholtz, *An Autobiographical Sketch: An Address Delivered on the Occasion of His Jubilee*, trans. E. Atkinson (London: Longmans, Green, 1898).

49. Cahan, *Life*, 51–4; Steege, *Helmholtz*, 66.
50. Qtd. in Steege, *Helmholtz*, 65–66.
51. Cahan, *Life*, 49–54; Steege, *Helmholtz*, 67–68.
52. Cahan, *Life*, 264–65; Picker, *Soundscapes*, 176.
53. Helmholtz, *Sensations*, 129.
54. Picker, *Soundscapes*, 86.
55. Helmholtz, *Sensations*, 129, 138.
56. Hermann von Helmholtz, *Popular Lectures on Scientific Subjects*, trans. E. Atkinson (New York: D. Appleton, 1881), 85.
57. Helmholtz, *Sensations*, 129.
58. Tyndall, *Sound*, 1st ed., 325.
59. Helmholtz, *Sensations*, 145, 148.
60. Steege, *Helmholtz*, 63.
61. Helmholtz, *Sensations*, 130, 7; Helmholtz, *Popular Lectures*, 61.
62. Kivy, *Music*, 17–21; Edmund Gurney, *The Power of Sound* (London: Smith, Elder, 1880), 103.
63. Gurney, *Power*, 167–68.
64. Chomet, *Influence*, 191.
65. Qtd. in Chomet, *Influence*, 192, emphasis in original.
66. "Some Physiological Effects of Music," *Stroud Journal* 39, no. 2056 (October 20, 1893): 8.
67. "Physiological Effects of Music," 8.
68. Rodolphe Radau, *Wonders of Acoustics: Or, The Phenomena of Sound* (New York: Charles Scribner's Sons, 1886), 236.
69. Havelock Ellis, *Studies in the Psychology of Sex*, vol. 4 (Philadelphia: F. A. Davis, 1905), 131.
70. Ellis, *Studies*, 4:131–32.
71. E. T. A. Hoffmann, "Beethoven's Instrumental Music," trans. Arthur Ware Locke, *Musical Quarterly* 3, no. 1 (January 1917): 127. Here, I refer to the Romantic movement in music, often termed German Romanticism. For more, see Johnson, *Listening in Paris*; James Donelan, *Poetry and the Romantic Musical Aesthetic* (Cambridge: Cambridge University Press, 2009); and Bowman, *Philosophical Perspectives*. For more on German idealist philosophy and music, see Bowman, *Philosophical Perspectives*; Christel Fricke, "Kant," in *Music in German Philosophy:*

An Introduction, ed. Stefan Lorenz Sorgner and Oliver Fürbeth, trans. Susan H. Gillespie (Chicago: University of Chicago Press, 2010), 33–34; Terry Eagleton, *The Ideology of the Aesthetic* (Oxford: Basil Blackwell, 1990).

72. James Kennaway, "Music and the Body in the History of Medicine," in Kim and Gilman, *Oxford Handbook of Music and the Body*, 338.

73. Hoffmann, "Beethoven's Instrumental Music," 127. Hoffmann's ideas were echoed by thinkers like Wilhelm Heinrich Wackendoder, Friedrich Schlegel, and Gustav Schilling. Walter Frisch, *Music in the Nineteenth Century* (New York: Norton, 2012), 17; Catherine Jones, *Literature and Music in the Atlantic World, 1767–1867* (Edinburgh: Edinburgh University Press, 2014), 166.

74. For more on absolute music and its aesthetic and ideological ramifications, see Donelan, *Poetry*; Carl Dalhaus, *The Idea of Absolute Music*, trans. Roger Lustig (Chicago: University of Chicago Press, 1989); Daniel Chua, *Absolute Music and the Construction of Meaning* (Cambridge: Cambridge University Press, 1999); Jackson, *Harmonious Triads*.

75. Alex Ross, "Hold Your Applause: Inventing and Reinventing the Classical Concert," Lecture at the Royal Philharmonic Society, March 8, 2010, *The Rest Is Noise* (blog), https://www.therestisnoise.com/2005/02/18/.

76. Jairo Moreno, "Body'n'Soul? Voice and Movement in Keith Jarrett's Pianism," *Musical Quarterly* 83, no. 1 (Spring 1999): 83.

77. Johnson, *Listening in Paris*, 240–42, 285. For more on the nineteenth-century attempts to "render audiences docile" in both music and theatre, see Johnson, *Listening in Paris*; Lawrence Levine, *Highbrow/Lowbrow: The Emergence of Cultural Hierarchy in America* (Cambridge, MA: Harvard University Press, 1988), 189; Sedgman, *Reasonable Audience*; Hannah Simpson, "Tics in the Theatre: The Quiet Audience, the Relaxed Performance, and the Neurodivergent Spectator," *Theatre Topics* 28, no. 3 (2018): 227–38.

78. "Concert Etiquette," *Dwight's Journal of Music* 24, no. 4 (May 14, 1864): 234.

79. Qtd. in Zon, *Music and Metaphor*, 15.

80. Matthew Wills, "Mary Somerville, Queen of 19th Century Science," *JSTOR Daily* (March 2, 2016), https://daily.jstor.org/mary-somerville-queen-of-19th-century-science/.

81. Mary Somerville to John Tyndall, December 27, 1873, RI MS JT/1/S/79, Royal Institution.

82. Nicholas Mitchell, "The Mystery of Music," in *London in Light and Darkness* (London: William Tegg, 1871), 233–35, lines 1–3.

83. Robert Michael Brain, *The Pulse of Modernism: Physiological Aesthetics in Fin-de-Siècle Europe* (Seattle: University of Washington Press, 2015), xv. For more on the contemporaneous discourse of psychophysics, which put more emphasis on cognitive perception, see Jonathan Crary, *Suspensions of Perception: Attention, Spectacle, and Modern Culture* (Cambridge, MA: MIT Press, 1999); Fretwell, *Sensory Experiments*.

84. Brain, *Pulse*, xv.

85. James Sully, *Sensation and Intuition: Studies in Psychology and Aesthetics* (London: Henry S. King, 1874), 38.

86. Dames suggests that physiologists fell into two categories: "soft" and "hard." "Soft" physiologists like Sully maintained that the intellect or higher-level mental functions played crucial roles in humans' aesthetic perceptions, while "hard" physiologists like Allen, Gurney, and Huxley emphasized "the nervous, or pre-cognitive, root of any sensation" and advanced the more monist, rationalist views that the universe operated according to mechanical principles and that human life was governed by material laws. Dames, *Physiology*, 146. Sully believed that sensation should be studied on the "borderland of physiology and psychology." Sully, *Sensation and Intuition*, 37–38.

87. Carolyn Burdett, " 'The Subjective Inside Us Can Turn into the Objective Outside': Vernon Lee's Psychological Aesthetics," *19: Interdisciplinary Studies in the Long Nineteenth Century* 12 (2011), https://19.bbk.ac.uk/article/id/1549/.

88. Burdett, "Subjective."

89. Burdett, "Subjective."

90. Vernon Lee, *Laurus Nobilis: Chapters on Art and Life* (London: John Lane, 1909), 80.

91. Acousticians and theorists of physiological aesthetics frequently corresponded. Herbert Spencer to John Tyndall, November 3, 1886, RI MS JT/1/S/129, Royal Institution; Spencer to Tyndall, March 16, 1887, RI MS JT/1/S/133, Royal Institution; T. H. Huxley to Tyndall, September 3, 1857, RI MS JT/1/H/518, Royal Institution; Huxley to Tyndall, July 11, 1877, RI MS JT/1/H/532, Royal Institution. Many of these men interacted at scientific organizations in London, including the Royal Institution, the Royal Society, and the X-Club. See Ruth Barton, " 'An Influential Set of Chaps': The X-Club and Royal Society Politics, 1865–85," *British Journal for the History of Science* 23, no. 1 (March 1990): 53–81; J. D. Burchfield, "John Tyndall at the Royal Institution," in *"The Common Purposes of Life": Science and Society at the Royal Institution of Great Britain*, ed. Frank A. J. L. James (London: Routledge, 2017), 147–68.

92. Grant Allen, *Physiological Aesthetics* (London: Henry S. King, 1877), 119.

93. Sully, *Sensation and Intuition*, 58, 169–70.

94. Vernon Lee, *Music and Its Lovers: An Empirical Study of Emotional and Imaginative Responses to Music* (New York: E. P. Dutton, 1933), 25, 95, 127, 152.

95. Lee, *Music*, 136.

96. Lee, *Music*, 243.

97. Morgan, *Outward*, 4; "Review: *Physiological Aesthetics* by Grant Allen," *Popular Science Monthly* 11 (October 1877): 760.

98. Qtd. in Morgan, *Outward*, 9.

99. Qtd. in David Trippett, *Wagner's Melodies: Aesthetics and Materialism in German Musical Identity* (Cambridge: Cambridge University Press, 2013), 396.

100. Qtd. in Steege, *Helmholtz*, 81.

101. Morgan, *Outward*, 11; Dames, *Physiology*, 95. As Phyllis Weliver points out, associationist psychologists like William Hamilton and E. S. Dallas had long theorized music making as an "automotive process[]." Weliver, *Women Musicians*, 8–9. Physiological science arose at a time when, Nicholas Dames writes, there was a "general attempt to downplay the role of alert cognition," and physical and physiological acoustics cast music appreciation as an automatic phenomenon. Dames, *Physiology*, 95–96.

102. Vernon Lee, *The Beautiful: An Introduction to Psychological Aesthetics* (Cambridge: Cambridge University Press, 1913), 24. For more on Lee's ambivalence about physiological aesthetics, see Burdett, "Subjective."

103. T. H. Huxley, "On the Hypothesis that Animals Are Automata, and Its History," *Fortnightly Review*, n.s., 16 (November 1, 1874): 577.

104. Max Nordau, *Degeneration* (London: William Heinemann, 1895), 56.

105. Nordau, *Degeneration*, 142.

106. Ernest Newman, "The World of Music: A Physiology of Criticism," *Sunday Times*, December 16, 1928: 7.

107. James Fennell, "On the Effects of Music on Man and Animals." *Mirror of Literature, Amusement, and Instruction* 38, no. 1081 (October 16, 1841): 245; James R. Tracy, "The Power of Music Over Animals," *Musical Standard*, August 11, 1894: 106–7; "The Influence of Music on the Lower Animals," *All the Year Round* 30, no. 735 (December 30, 1882): 538; Chomet, *Influence*, 190.

108. Ellis, *Studies*, vol. 4, 122.

109. Edmund Gurney, "On Some Disputed Points in Music," *Fortnightly Review* 20, no. 115 (1876): 128–30.

110. Gurney, *Power*, 31.

111. Hui, *Psychophysical Ear*, 85; Fretwell, *Sensory Experiments*, 91, 110; Zon, *Evolution*, 84–85.

112. Qtd. in Trippett, *Wagner's Melodies*, 337–38.

113. Ellis, *Studies*, vol. 4, 122.

114. "Musical Vibrations," *Weston-super-Mare Gazette, and General Advertiser* 41, no. 2210 (October 21, 1885): 3.

115. E. Carey, "Body and Music," *National Review* 5, no. 27 (May 1885): 382–90.

116. Carey, 384–85.

117. Carey, 385, 390.

118. Cahan, "Elite," 55; Gillin, *Sound Authorities*, 17; Steege, *Helmholtz*, 24.

119. Qtd. in Russell Kahl, introduction to *The Selected Writings of Hermann von Helmholtz*, by Hermann von Helmholtz, ed. Russell Kahl (Middletown, CT: Wesleyan University Press, 1971), xii–xiii.

120. Steege, *Helmholtz*, 6.

121. "Dr. Swinderton Heap," *Musical Herald* 547 (October 2, 1893): 293.

122. "Voice Failure and Its Attendant Ailments," *Musical Herald*, no. 619 (October 1, 1899): 309.

123. Qtd. in Edward Baughan, "Marchesi and Singing," *Monthly Musical Record* 28, no. 325 (1898): 7–8.

124. Baughan, "Marchesi," 7–8.

125. "The Organ World: On the Physiology of Pianoforte Playing, with a Practical Application of a New Theory," *Musical World* 68, no. 3 (November 3, 1888): 849.

126. Jackson, *Harmonious Triads*, 234–35.

127. Sedley Taylor, *Sound and Music: A Non-Mathematical Treatise on the Physical Constitution of Musical Sounds and Harmony, including The Chief Acoustical Discoveries of Professor Helmholtz* (London: Macmillan, 1883), vii. The bookplate on the inside of the edition of Taylor's book that is currently in the Royal College of Music Library reveals that the book was presented to the Royal College of Music by Macmillan & Co. on October 17, 1883. Elfrieda Hiebert, "Listening to the Piano Pedal: Acoustics and Pedagogy in Late Nineteenth-Century Contexts," *Osiris* 28, no. 1 (January 2013): 240; T. F. Harris, *Hand Book of Acoustics for the Use of Musical Students*, 3rd ed. (London: J. Curwen and Sons, 1887), iii–v.

128. John Broadhouse, *The Student's Helmholtz: Musical Acoustics; or, The Phenomena of Sound as Connected with Music* (London: William Reeves, 1881), vii–ix; advertisement for degree of doctor of music at Oxford University, *Musical News* (September 4, 1891): 543, in MS, Royal College of Music Archives, Royal College of Music Library, London, accessed July 5, 2017; advertisement for Trinity College Fellowship Examination, *Musical News* (March 6, 1891): 13, in MS, Royal College of Music Archives, Royal College of Music Library, London, accessed July 5, 2017.

129. Advertisement for degree of doctor of music at Oxford University.

130. "Novello, Ewer, & Co.'s Music Primers" (advertisement), *Musical Times and Singing Class Circular* 21, no. 451 (September 1, 1880): 470.

131. W. H. Stone, *The Scientific Basis of Music* (London: Novello, Ewer, 1878).

132. Allaston Burgh, *Anecdotes of Music, Historical and Biographical; in a Series of Letters from a Gentleman to His Daughter* (London: Longman, Hurst, Rees, Orme, and Brown, 1814), 10–11; William Davis, "Music Therapy in Victorian England," *British Journal of Music Therapy* 2, no. 1 (1988): 10–16; Jillian C. Rogers, *Resonant Recoveries: French Music and Trauma Between the World Wars* (Oxford: Oxford University Press, 2021); Rosemary Golding, *Music and Moral Management in the Nineteenth-Century English Lunatic Asylum* (London: Palgrave Macmillan, 2021).

133. "Music as a Relief to Pain," *Musical Times and Singing-Class Circular* 22, no. 463 (September 1, 1881): 458.

134. Henry C. Lunn, "Musical Doctors," *Musical Times and Singing-Class Circular* 21, no. 452 (October 1, 1880): 495–96.

135. J. Ewing Hunter, "Is Soft Music a Calmative in Cases of Fever?" *British Medical Journal* 2, no. 1660 (1892): 923.

136. Davis, "Music Therapy," 11–12.
137. Davis, 11–12.
138. "Medicinal Music," *Musical Times and Singing Class Circular* 32, no. 584 (October 1, 1891): 587–88.
139. R. H. McCartney, *Barbell or Wand Exercises for Use in Schools with Musical Accompaniment* (London: George Gill and Sons, 1881), 32.
140. Flora T. Parsons, *Callisthenic Songs Illustrated: A New and Attractive Collection of Callisthenic Songs Beautifully Illustrated* (New York: Ivison, Blakeman, Taylor, 1869), v.
141. McGuire, *Music*, 122–24; Zon, *Evolution*, 169–75; Steege, *Helmholtz*, 198. Helmholtz attended a meeting of the Tonic Sol-Fa organization on one of his London trips. Zon, *Evolution*, 203–5.

Chapter Two

1. Burgan, "Heroines," 51.
2. Burgan, "Heroines," 51; Gillett, *Musical Women*, 77–78.
3. Gillett, *Musical Women*, 77–78.
4. Ward, *Robert Elsmere*, 253.
5. Gillett, *Musical Women*, 98, 140.
6. William Crawford Honeyman, *The Secrets of Violin Playing* (Edinburgh: E. Köhler and Son, 1800s?), 74.
7. Lorde, "Uses of the Erotic," 54–55.
8. Gillett, *Musical Women*, 4.
9. George Eliot, *Daniel Deronda*, ed. Terence Cave (London: Penguin, 1995), 52.
10. Weliver, *Women Musicians*, 111–12; Delia da Sousa Correa, *George Eliot, Music and Victorian Culture* (London: Palgrave Macmillan, 2003), 81–84.
11. Weliver, *Women Musicians*, 111–12.
12. Emily Auerbach, *Maestros, Dilettantes, and Philistines: The Musician in the Victorian Novel* (Bern, Switzerland: Peter Lang, 1989), 178; Rebecca Pope, "The Diva Doesn't Die: George Eliot's 'Armgart,'" *Criticism* 32, no. 4 (Fall 1990): 469–70.
13. For further discussion of these novels, see Gillett, *Musical Women*.
14. Daisy Rhodes Campbell, *The Violin Lady* (Boston: Page Company, 1916), 322.
15. Marion Scott, "British Women as Instrumentalists," *Music Student* 10, no. 9 (May 1913): 338; Gillett, *Musical Women*, 67.
16. Edith Lynwood Winn, "The Study of the Violin for Girls," *Musical World* 1, no. 10 (November 1901): 132.
17. Jane Bowers and Judith Tick, introduction to *Women Making Music: The Western Art Tradition, 1150–1950*, ed. Jane Bowers and Judith Tick (Urbana:

University of Illinois Press, 1986), 7–8; Ada Molteno, "Ladies as Orchestral Players By One of Them," *Orchestral Association Gazette*, no. 6 (March 1894): 63.

18. Caroline Blanche Elizabeth Lindsay, "How to Play the Violin," in *The Girl's Own Indoor Book*, ed. Charles Peters (Philadelphia: J. B. Lippincott, 1892), 174.

19. Scott, "British Women," 337.

20. F. G. E., "Lady Violinists (Concluded)," *Musical Times* 47, no. 765 (November 1, 1906): 739.

21. Molteno, "Ladies," 63. For more on ladies' orchestras, see Shannon Draucker, "Ladies' Orchestras and Music-As-Performance in *Fin-de-Siècle* Britain," *Nineteenth-Century Contexts* 45, no. 1 (2023): 7–22.

22. Lucy Green, *Music, Gender, Education* (Cambridge: Cambridge University Press, 1997), 67.

23. Richard Morrison, *Orchestra: The LSO; A Century of Triumph and Turbulence* (London: Faber and Faber, 2004), 186–87.

24. Winn, "Study of the Violin," 132.

25. Qtd. in Wallace Sutcliffe, "Ladies as Orchestral Players," *Orchestral Association Gazette* 5 (February 1894): 49.

26. Lindsay, "How to Play," 174.

27. Gillett, *Musical Women*, 87.

28. Gillett, *Musical Women*, 6.

29. See William G. Weber, *Music and the Middle Class: The Social Structure of Concert Life in London, Paris, and Vienna between 1830 and 1848* (New York: Holmes and Meier, 1975).

30. Gillett, *Musical Women*, 7; Derek B. Scott, *From the Erotic to the Demonic: On Critical Musicology* (Oxford: Oxford University Press, 2003), 36.

31. Gillett, *Musical Women*, 111; "The Fair Sex-Tett," *Punch* 68 (April 3, 1875): 150.

32. Edmund S. J. van der Straeten, *The Technics of Violoncello Playing* (London: The Strad, 1898), 18.

33. Gillett, *Musical Women*, 101. See also Tilden Russell, "The Development of the Cello Endpin," *Imago Musicae* 4 (1985): 335–56; Tilden Russell, "Endpin," *Grove Music Online, Oxford Music Online* (Oxford: Oxford University Press, 2001), https://doi.org/10.1093/gmo/9781561592630.article.08788.

34. Gillett, *Musical Women*, 4.

35. "Joachim's Rival," *Musical World* 10, no. 58 (March 6, 1880): 153.

36. "Philharmonic Concerts," *Athenaeum: Journal of Literature, Science, and the Fine Arts* 347, no. 21 (June 1834): 475.

37. William T. Parke, *Musical Memoirs; Comprising an Account of the General State of Music in England, from the First Commemoration of Handel, in 1784, to the year 1830* (London: Henry Colburn and Richard Bentley, 1830), 130.

38. Gillett, *Musical Women*, 18.

39. Sutcliffe, "Ladies as Orchestral Players," 48–49.

40. Honeyman, *Secrets*, 74.

41. Honeyman, 74.

42. Society of Women Musicians to Sir John Reight, Director General, British Broadcasting Company (October 30, 1928), MS, MS. Society of Women Musicians Papers (Box 4), Royal College of Music Library, London.

43. Society of Women Musicians to Sir John Reight.

44. Lunn, "Musical Doctors," 496.

45. Qtd. in Cheng, *Loving*, 144.

46. Qtd. in Sutcliffe, "Ladies as Orchestral Players," 49.

47. Florence G. Fidler, "Women as Orchestral Players," *Musical News* 38, no. 474 (March 31, 1900): 310.

48. "Women in Orchestras," *Musical Standard* 46 (January 20, 1894): 47.

49. Molteno, "Ladies," 64; Lindsay, "How to Play," 176.

50. "An Evening with the Royal Amateur Orchestral Society," *Musical Standard* 34, no. 19 (February 1888): 98.

51. Charles Barnard, *Camilla, The Tale of a Violin, being the Artist Life of Camilla Urso* (Boston: Loring, 1874), 15, 17–18.

52. Barnard, 28.

53. Miriam Elizabeth Burstein, introduction to Ward, *Robert Elsmere*, 6.

54. Valerie Sanders, *Eve's Renegade: Victorian Anti-Feminist Women Novelists* (New York: Palgrave Macmillan, 1996); J. Russell Perkin, *Theology and the Victorian Novel* (Montreal: McGill University Press, 2009); Judith Wilt, "The Romance of Faith: Mary Ward's *Robert Elsmere* and *Richard Meynell*," *Literature and Theology* 10, no. 1 (March 1996): 33–43.

55. Janet Trevelyan, *The Life of Mrs. Humphry Ward* (New York: Dodd, Mead, 1923), 21, 127.

56. Weliver, *Musical Crowd*, 164.

57. For discussions of Rose as passionate and subversive, see Perkin, *Theology*; Gisela Argyle, "Mrs. Humphry Ward's Fictional Experiments in the Woman Question," *Studies in English Literature, 1500–1900* 43, no. 4 (Autumn 2003): 939–57. For lamentations of Rose's marriage, see Gillett, *Musical Women*; Sanders, *Eve's Renegade*; Burstein, introduction. Gillett and Weliver are among the few critics to discuss Rose's musicianship. Gillett mentions Rose's violin playing but does not analyze the novel at length. Gillett, *Musical Women*, 79. Weliver argues that music's role in the novel is to reflect "notions of national spirit or identity," critique the "moral dangers and affectations of high society," and add "aesthetic purpose" to a work of "rational-critical" fiction otherwise concerned with a "dry theological debate." Weliver, *Musical Crowd*, 157–59.

58. Ward, *Robert Elsmere*, 247, 282, 177; Trevelyan, *Life*, 79; Thomas Ward, Family Diaries, 1886, 1887, 1888, 1892, MS 202/5A, Mary Ward Papers, University College London Archives, National Archives, London.

59. Ward, *Robert Elsmere*, 28.

60. Scott, *Singing Bourgeois*, 12.

61. Clive Brown, "Louis Spohr," *Grove Music Online, Oxford Music Online* (Oxford: Oxford University Press, 2001), https://www.oxfordmusiconline.com/grovemusic/view/10.1093/gmo/9781561592630.001.0001/omo-9781561592630-e-0000026446; Ward, *Robert Elsmere*, 28. For more on the types of music Rose plays and how these musical references resonate with Matthew Arnold's ideas about progress and modernity, see Weliver, *Musical Crowd*.

62. Ward, *Robert Elsmere*, 55.

63. Ward, 54–55.

64. Ward, 252.

65. Ward, 175.

66. Ward, 316.

67. Ward, 54–55.

68. Ward, 106.

69. Ward, 253.

70. Theodor Billroth, *Wer ist musikalisch?* (Berlin: Gebrüder Paetel, 1895); Neil McLaren and Rafael Vara Thorbeck, "Little-Known Aspect of Theodor Billroth's Work: His Contribution to Musical Theory," *World Journal of Surgery* 21 (1997): 569–71.

71. Franz Joseph Gall, *On the Functions of the Brain and Each of Its Parts*, vol. 1, trans. Winslow Lewis (Boston: Marsh, Capen, and Lyon, 1835), 107, 245; Céline Frigau Manning, "Phrenologizing Opera Singers: The Scientific 'Proofs of Musical Genius,'" *19th Century Music* 39, no. 2 (Fall 2015): 125–41; Paul Eling, Stanley Finger, and Harry Whitaker, "On the Origins of Organology: Franz Joseph Gall and a Girl Named Bianchi," *Cortext* 86 (2017): 123–31; Alan Davison, "High-Art Music and Low-Brow Types: Physiognomy and Nineteenth-Century Music Iconography," *Context* 17 (Winter 1999): 5.

72. François Joseph Gall, *On the Functions of the Brain and Each of Its Parts*, vol. 5, trans. Winslow Lewis (Boston: Marsh, Capen, and Lyon, 1835), 63; Manning, "Phrenologizing."

73. Carl Seashore, "The Measurement of Musical Talent," *Musical Quarterly* 1, no. 1 (1915): 129–48.

74. Ward, *Robert Elsmere*, 236.

75. Tyndall, *Sound*, 1st ed., 200.

76. Ward, *Robert Elsmere*, 189–90.

77. Tyndall, *Sound*, 1st ed., 49.

78. Ward, *Robert Elsmere*, 253.

79. Ward, 459.

80. Ward, 251.

81. Ward, 251–53.

82. For a reading of this scene in the context of Victorian crowd theory, see Weliver, *Musical Crowd*, 167.

83. Ward, *Robert Elsmere*, 397.
84. Ward, 55.
85. Ward, 55.
86. Ward, 396.
87. Lindsay, "How to Play," 175.
88. Henry Saint-George, "The Bow, Its History, Manufacture and Use," *The Strad* 7, no. 78 (October 1896): 175.
89. George Dubourg, *The Violin: Some Account of That Leading Instrument and Its Most Eminent Professors* (London: Robert Cocks, 1836), 172.
90. Anna Leffler Arnim, *A Complete Course of Wrist and Finger Gymnastic: For Students of the Piano, Organ, Violin, and Other Instruments*, 3rd ed. (London: Hutchings and Crowsley, 1894), 13.
91. Ward, *Robert Elsmere*, 55.
92. Ward, 106–7.
93. Ward, 107.
94. Ward, 190, 394, 254.
95. Ward, 396–97.
96. Ward, 253.
97. Both Gillett and Weliver read Rose's marriage to Hugh as evidence of Ward's disappointing gender politics. Gillett argues that Rose's marriage to Hugh "implies the irrelevance of her past ambitions," and Weliver considers the moment as reflective of Ward's "conservative views regarding women." Gillett, *Musical Women*, 119; Weliver, *Musical Crowd*, 157.
98. Mary Augusta Ward, *The Case of Richard Meynell* (Garden City, NY: Doubleday, 1911), 49.
99. Francis, *Duenna*, 5.
100. Francis, 5.
101. Francis, 9.
102. Francis, 18, 49.
103. Francis, 44.
104. Francis, 45.
105. Francis, 166.
106. Francis, 138–39.
107. Francis, 138–39.
108. Francis, 223.
109. Francis, 60–61.
110. Francis, 49.
111. Francis, 45.
112. Francis, 49.
113. Francis, 49.
114. Francis, 61.
115. Francis, 45.

116. Jackson Beatty and Brennis Lucero-Wagoner, "The Pupillary System," in *Handbook of Psychophysiology*, ed. John T. Cacioppo, Louis G. Tassinary, and Gary G. Berntson (Cambridge: Cambridge University Press, 2000), 143.

117. Qtd. in Beatty and Lucero-Wagoner, "Pupillary," 143.

118. Francis, *Duenna*, 114.

119. George Henry Lewes, *The Physiology of Common Life*, vol. 2 (New York: D. Appleton, 1875), 56. As Weliver notes, eighteenth-century associationist scientists like David Hartley used instrument playing as a representative example of an "automotive process[]": "through repeated practice the process becomes so automatic that conscious thought is no longer required." Weliver, *Women Musicians*, ch. 2, appendix 1; David Hartley, *Observations on Man, His Frame, His Duty, and His Expectations*, 6th ed. (London: Thomas Tegg and Son, 1834), 69.

120. H. Hayes Newington, "Some Mental Aspects of Music," *Journal of Mental Science* 43 (October 1897): 712.

121. Newington, "Some Mental Aspects," 713.

122. Herbert Spencer, *The Principles of Psychology*, 2nd ed. (London: Williams and Norgate, 1870), 451.

123. Francis, *Duenna*, 234.

124. Francis, 236.

125. Francis, 243.

126. Francis, 249.

127. Francis, 249.

128. Francis, 244.

129. Francis, 227–28.

130. Francis, 227.

131. Francis, 233.

132. Francis, 233.

133. Francis, 253.

134. Francis, 255.

135. Gillett, *Musical Women*, 124.

136. Francis, *Duenna*, 254.

137. Francis, 253.

138. Przybylo, *Asexual Erotics*, 22.

139. Francis, *Duenna*, 259.

140. Francis, 274.

141. Francis, 257.

142. Francis, 209.

143. Francis, 277–78.

144. Francis, 279.

145. Francis, 278.

146. Elizabeth A. Wilson, *Gut Feminism* (Durham, NC: Duke University Press, 2015), 24.

147. Wilson, *Psychosomatic*, 13–14; Grosz, *Volatile Bodies*, ix; Willey, "Biopossibility," 510.

148. Elizabeth Grosz, *Time Travels: Feminism, Nature, Power* (Durham, NC: Duke University Press, 2005), 4.

Chapter Three

1. Bertha Thomas, *The Violin-Player* (London: Richard Bentley and Son, 1880), I.199.

2. Thomas, I.103.

3. Gillett, *Musical Women*, 78.

4. Thomas, *Violin-Player*, I.104, 107.

5. Thomas, I.153.

6. Judith Butler, *Gender Trouble: Feminism and the Subversion of Identity* (New York: Routledge, 2006), 190–92.

7. Sarah Grand, *The Heavenly Twins*, edited by Carol Senf (Ann Arbor: University of Michigan Press, 1994), 452.

8. Thomas, *Violin-Player*, I.107.

9. Responding to critiques by theorists such as Martha Nussbaum that *Gender Trouble* failed to account for the realities of the body, Butler revised her discussion of gender essentialism in *Bodies That Matter* (1993): "Bodies live and eat; eat and sleep; feel pain and pleasure; endure illness and violence . . . these facts . . . cannot be dismissed as mere construction." Butler, *Bodies That Matter: On the Discursive Limits of "Sex"* (New York: Routledge, 2011), viii.

10. See Richardson, *Love and Eugenics*, chs. 5 and 6.

11. For recent debates on the terms "pregnant person" versus "pregnant woman," see Michael Powell, "A Vanishing Word in the Abortion Debate: 'Woman,'" *New York Times*, June 8, 2022, https://www.nytimes.com/2022/06/08/us/women-gender-aclu-abortion.html; Helen Lewis, "Why I'll Keep Saying 'Pregnant Women,'" *The Atlantic*, October 26, 2021, https://www.theatlantic.com/ideas/archive/2021/10/pregnant-women-people-feminism-language/620468/; Shannon Palus, "How to Think About the Debate Over the Phrase 'Pregnant People,'" *Slate*, July 9, 2022, https://slate.com/technology/2022/07/pregnant-people-inclusive-language-gender-debate.html. I am grateful to Doreen Thierauf for bringing these pieces to my attention.

12. Przybylo, *Asexual Erotics*, 20. See also Prosser, *Second Skins*, and Stryker, "Transgender Studies."

13. Taylor, "Taking It in the Ear," 612.

14. Baitz, "Toward a Trans* Method," 374–76.

15. Clare Taylor, *Women, Writing, and Fetishism, 1890–1950: Female Cross-Gendering* (Oxford: Oxford University Press, 2003), 24–25.

16. Clare Taylor, for instance, writes that the story "dismembers" the novel "both generically and thematically." Taylor, *Women*, 24–25.

17. Qtd. in Taylor, 24.

18. See Taylor, *Women*; Carol Senf, introduction to Grand, *Heavenly Twins*, vii–xxxvii; John Kucich, "Curious Dualities: *The Heavenly Twins* (1893) and Sarah Grand's Belated Modernist Aesthetics," in *The New Nineteenth Century: Feminist Readings of Underread Victorian Fiction*, ed. Barbara Leah Harman and Susan Meyer (New York: Routledge, 1996), 195–204.

19. Grand, *Heavenly Twins*, 61.

20. Grand, 450.

21. Martha Vicinus, "Turn-of-the-Century Male Impersonation: Rewriting the Romance Plot," in *Sexualities in Victorian Britain*, ed. Andrew H. Miller and James Eli Adams (Bloomington: Indiana University Press, 1996), 204.

22. Grand, *Heavenly Twins*, 189.

23. Grand, 437.

24. Grand, 392.

25. Wendy Bashant, "Singing in Greek Drag: Gluck, Berlioz, George Eliot," in *En Travesti: Women, Gender Subversion, Opera*, ed. Corinne E. Blackmer and Patricia Juliana Smith (New York: Columbia University Press, 1995), 234.

26. Grand, *Heavenly Twins*, 399.

27. Grand, 437.

28. See Bernstein, "Music Framed"; McGann, *Poetics of Sensibility*.

29. Grand, *Heavenly Twins*, 403.

30. Grand, 403.

31. Quoting Percy Shelley's 1820 poem "The Witch of Atlas," the Tenor remarks that the Boy has developed "no defect / Of either sex, yet all the grace of both." Grand, 403.

32. Grand, 446.

33. Grand, 446–47.

34. Grand, 458.

35. Grand, 456.

36. Grand, 471–72.

37. Grand, 472.

38. Grand, 476.

39. Grand, 477–78.

40. Grand, 541.

41. Grand, 453; Gillett, *Musical Women*, 33.

42. Ann Heilmann, *New Woman Strategies: Sarah Grand, Olive Schreiner, and Mona Caird* (Manchester: Manchester University Press, 2004), 65.

43. Grand, *Heavenly Twins*, 480.

44. Grand, 1.

45. Richardson, *Love and Eugenics*, 142.

46. Senf, introduction, ix.

47. Nigel Burton, "Florence Ashton Marshall," in *Grove Music Online, Oxford Music Online* (Oxford: Oxford University Press, 2001), https://doi.org/10.1093/gmo/9781561592630.article.2020251.

48. Burton, "Florence"; "Prince Sprite," *Musical Times* 33, no. 593 (July 1, 1892): 441.

49. I am grateful to Sophie Fuller for bringing Frances Thomas and her career as a clarinetist to my attention at the Music in Nineteenth-Century Britain Conference in June 2017.

50. Pamela Weston, *Clarinet Virtuosi of the Past* (London: Robert Hale, 1971), 257. These awards included a commendation for harmony, a silver medal for clarinet, and a certificate of merit for clarinet. "Royal Academy of Music," *Musical Times* 17, no. 390 (August 1, 1875): 170–71; "Royal Academy of Music: Female Department," *Musical Times* 18, no. 414 (August 1, 1877): 390; "Music," *Illustrated London News* 69, no. 1931 (July 29, 1876): 114–15.

51. Weston, *Clarinet Virtuosi*, 257–58.

52. *Musical Times and Singing-Class Circular* 35, no. 617 (July 1, 1894): 482; *Musical Times and Singing-Class Circular* 37, no. 638 (April 1, 1896): 266; "The Musical Artists' Society," *Musical Times* 34, no. 604 (June 1, 1893): 346; Weston, *Clarinet Virtuosi*, 257–58.

53. "Miscellaneous Concerts," *Musical World* 65, no. 29 (June 16, 1887): 559.

54. "Review: *The Violin-Player* by Bertha Thomas," *The Graphic* (July 31, 1880): 119.

55. Thomas, *Violin-Player*, I.3.

56. Thomas, I.5–6.

57. Thomas, I.8.

58. Billroth, *Wer ist musikalisch?*; Manning, "Phrenologizing," 133.

59. Thomas, *Violin-Player*, I.24.

60. Thomas, I.28.

61. Thomas, I.28.

62. See Daniel Levitin, *This Is Your Brain on Music: The Science of a Human Obsession* (New York: Penguin, 2006); Oliver Sacks, *Musicophilia: Tales of Music and the Brain*, rev. and exp. ed. (New York: Vintage, 2007).

63. R. H. M. Bosanquet, *An Elementary Treatise on Musical Intervals and Temperament* (London: Macmillan, 1876), xiv–xv.

64. "Sense of Absolute Pitch," *Musical Herald*, no. 577 (April 1, 1896): 108–9.

65. Thomas, *Violin-Player*, I.116.

66. Dubourg, *Violin*, 107, 246.

67. Helmholtz, *Sensations*, 116, 120.

68. Thomas, *Violin-Player*, I.116–17.

69. Thomas, I.199–200.

70. Thomas, I.266.

71. Thomas, II.218.
72. Thomas, II.219.
73. Thomas, I.278.
74. Thomas, III.241.
75. Thomas, II.34–36.
76. Thomas, II.34–36.
77. Thomas, I.128.
78. David Tod, *The Anatomy and Physiology of the Organ of Hearing* (London: Longman, Rees, Orme, Brown, Green, and Longman, 1832), 58.
79. William Hyde Wollaston, "On Sounds Inaudible by Certain Ears," *Proceedings of the Royal Society of London, Philosophical Transactions of the Royal Society* (1820), 314.
80. Thomas, *Violin-Player*, III.93–94.
81. Thomas, III.93–94.
82. Thomas, III.23.
83. George Upton, *Musical Memories* (Chicago: A. C. McClurg, 1908), 70–71.
84. Thomas, *Violin-Player*, III.93.
85. Thomas, III.98.
86. Thomas, III.180.
87. Thomas, II.220. Laurence refuses several marriage proposals, including one from Gervase. After Gervase dies, she refuses a proposal from a "second-rate baritone's deputy" named Tristan, explaining, "I cherish my liberty. I should not make a good wife to you. I see now I am never to live but for music." Thomas, III.7, 52.
88. Thomas, III.104.
89. Thomas, III.219.
90. In an odd sensationalist moment, Linda's brother Bruno Pagano, against whom Gervase once brought charges of robbery, pushes Gervase off a cliff. Thomas, III.320.
91. Thomas, III.299. Though there is some insinuation that Val wishes to become Laurence's romantic partner, in the final scene, he sees her hand resting on her violin case and realizes that "in there lay her only life-companion. He understood." Thomas, III.319.
92. Butler, *Gender Trouble*, 195.
93. "Female Orchestral Players," *Musical News* 4, no. 112 (April 22, 1893): 372.
94. Fidler, "Women," 310.

Chapter Four

1. George Eliot, *Romola*, ed. Dorothea Barrett (London: Penguin Books, 1996), 85–86.

2. William Cheng, *Just Vibrations: The Purpose of Sounding Good* (Ann Arbor: University of Michigan Press, 2016), 73–75.

3. Suzanne Cusick, "Musicology, Torture, Repair," *Radical Musicology* 3 (2008), http://www.radical-musicology.org.uk/2008/Cusick.htm.

4. This was also true, as Jenny Olivia Johnson writes, in cases such as that of the Columbus Boychoir School in Princeton, New Jersey, in which "the 'mystical' power of music is alleged to have been used to seduce young students." Jenny Olivia Johnson, "Musical Abjects: Sounds and Objectionable Sexualities," in Maus and Whiteley, *Oxford Handbook of Music and Queerness*, 406.

5. See Przybylo, *Asexual Erotics*; Willey, *Undoing Monogamy*; Taylor, "Sound Desires."

6. Lana Dalley and Kellie Holzer, "Victorian Literature in the Age of #MeToo: An Introduction," *Nineteenth-Century Gender Studies* 16, no. 2 (Summer 2020), http://ncgsjournal.com/issue162/introduction.html.

7. Qtd. in Elissa Gurman, "Sex, Consent, and the Unconscious Female Body: Reading *Tess of the d'Urbervilles* alongside the Trial of Brock Allen Turner," *Law and Literature* 32, no. 1 (2019): 158.

8. Spampinato, "Rereading Rape," 142.

9. Linda Martín Alcoff, *Rape and Resistance* (Cambridge, MA: Polity, 2018), 2, 9.

10. Martín Alcoff, 77.

11. Martín Alcoff, 12.

12. Spampinato, "Rereading Rape," 127–28.

13. Spampinato, 145.

14. Martín Alcoff, *Rape*, 13.

15. Spampinato, "Rereading Rape," 138.

16. Kimberly Cox, *Touch, Sexuality, and Hands in British Literature, 1740–1901* (London: Routledge, 2022), 20, 54.

17. Kathleen Lubey, "Sexual Remembrance in *Clarissa*," *Eighteenth-Century Fiction* 29, no. 1 (Winter 2016–17): 157.

18. For more on the critical history of the debates about rape in *Tess*, see Spampinato, "Rereading Rape," 136–41; Gurman, "Sex, Consent," 161–62.

19. Thierauf, "Marital Rape."

20. Taylor, "Sound Desires," 277.

21. Spampinato, "Rereading Rape," 144.

22. Martín Alcoff, *Rape*, 77; Barbara Johnson, "Muteness Envy," in *The Barbara Johnson Reader: The Surprise of Otherness*, ed. Melissa Feuerstein et al. (Durham, NC: Duke University Press, 2014): 214, qtd. in Spampinato, "Rereading Rape."

23. Catharine MacKinnon, *Toward a Feminist Theory of the State* (Cambridge, MA: Harvard University Press, 1989), 148. For more on the phenomenon of physiological arousal during unwanted sex, see Meredith Chivers and Lori Brotto,

"Controversies of Women's Sexual Arousal and Desire," *European Psychologist* 22, no. 1 (2017): 5–26; Natasha McKeever, "Can a Woman Rape a Man and Why Does It Matter?," *Criminal Law and Philosophy* 13 (2019): 599–619.

24. Martín Alcoff, *Rape*, 114.

25. Martín Alcoff, 2, 114.

26. Kate Manne, *Down Girl: The Logic of Misogyny* (Oxford: Oxford University Press, 2018), 131.

27. Manne, 197.

28. Manne, 197.

29. See Thierauf, "Marital Rape," 265.

30. See Beryl Gray, *George Eliot and Music* (New York: Palgrave Macmillan, 1989); da Sousa Correa, *George Eliot*.

31. Gray, *George Eliot*, 15.

32. George Eliot, "Liszt, Wagner, and Weimar," *Fraser's Magazine* 52 (July 1855): 48–62.

33. For more on Eliot and music, see Alison Byerly, *Realism, Representation, and the Arts in Nineteenth-Century Literature* (Cambridge: Cambridge University Press, 1997), 133–36; Weliver, *Women Musicians*, 155; Auerbach, *Maestros*, 139; Shirley Frank Levenson, "The Use of Music in *Daniel Deronda*," *Nineteenth-Century Fiction* 24, no. 3 (1969): 319; Delia da Sousa Correa, "Music," in *George Eliot in Context*, ed. Margaret Harris (Cambridge: Cambridge University Press, 2013), 206; Rosemary Ashton, *George Eliot, a Life* (London: Penguin, 1997), 276; Oliver Lovesey, "Religion," in Harris, *George Eliot in Context*, 238. For more on music and sympathy in Eliot's works, see Gray, *George Eliot*, x; da Sousa Correa, "Music," 213. For more on Eliot's ethics of sympathy more broadly, see Rae Greiner, *Sympathetic Realism in Nineteenth-Century British Fiction* (Baltimore: Johns Hopkins University Press, 2012); Rachel Ablow, *The Marriage of Minds: Reading Sympathy in the Victorian Marriage Plot* (Stanford, CA: Stanford University Press, 2007).

34. George Eliot, "O May I Join the Choir Invisible," in *The Legend of Jubal and Other Poems* (Boston: James Osgood, 1874), 233; Nancy Henry, *The Life of George Eliot: A Critical Biography* (West Sussex: Wiley Blackwell, 2012), 182.

35. Picker, *Soundscapes*, 88; da Sousa Correa, *George Eliot*, 33–43.

36. Picker, *Soundscapes*, 87.

37. Eliot, *Journals*, 135; Picker, *Soundscapes*, 87–91; Weliver, *Women Musicians*, 194–95; da Sousa Correa, *George Eliot*, 33–43, 113–14.

38. George Eliot, "Mr. Gilfil's Love Story," in *Scenes of Clerical Life*, ed. Jennifer Gribble (London: Penguin Books, 1998), 189–91.

39. Eliot, *Daniel Deronda*, 60–61.

40. Eliot, 60–61; Picker, *Soundscapes*, 93; George Eliot, *The George Eliot Letters*, ed. Gordon S. Haight, vol. 6 (New Haven, CT: Yale University Press, 1954), 304. For more on this scene, see Gillian Beer, *George Eliot* (Brighton, UK: Harvester Press, 1986), 224; Doreen Thierauf, "Tending to Old Stories: *Daniel Deronda* and

Hysteria, Revisited," *Victorian Literature and Culture* 46, no. 2 (June 2018): 456; George Eliot, *George Eliot's Daniel Deronda Notebooks*, ed. Jane Irwin (Cambridge: Cambridge University Press, 1996): 351; Clapp-Itnyre, *Angelic Airs*, 124.

41. Weliver, *Women Musicians*, 213, 189.
42. da Sousa Correa, *George Eliot*, 123, 128.
43. Burdett, "Sexual Selection," 33; Picker, *Soundscapes*, 89.
44. Picker, *Soundscapes*, 89.
45. Gray, *George Eliot*, 57.
46. Weliver, *Women Musicians*, 200, emphasis in original.
47. Dalley and Holzer, "Victorian Literature."
48. Vanessa Ryan, *Thinking Without Thinking in the Victorian Novel* (Baltimore: Johns Hopkins University Press, 2012), 67.
49. George Eliot, *The Mill on the Floss*, ed. A. S. Byatt (London: Penguin Books, 2003), 484, 487, 497, 486.
50. See Thierauf, "Marital Rape," 265.
51. Eliot, *Mill*, 400, 297–98.
52. Eliot, 400.
53. Eliot, 401.
54. Eliot, 399–400.
55. Eliot, 401.
56. Eliot, 400.
57. Eliot, 400.
58. Eliot, 433–34.
59. Elisha Cohn, *Still Life: Suspended Development in the Victorian Novel* (Oxford: Oxford University Press, 2016), 95. Ryan also discusses the role of "involuntary action" in *The Mill on the Floss*, given Maggie's many "dream state[s], guided by unconscious, reflexive processes." Ryan, *Thinking*, 67.
60. Eliot, *Mill*, 435.
61. Eliot, 435–36.
62. Eliot, 458–59.
63. Allen, *Physiological Aesthetics*, 114–15.
64. Allen, 115.
65. Frederick Charles Baker, *How We Hear: A Treatise on the Phenomena of Sound* (London: Vincent Music, 1901), 84.
66. Eliot, *Mill*, 458–59.
67. Eliot, 460–61.
68. Eliot, 460–61.
69. For more on the violence of acts of hand-grabbing, see Cox, *Touch*.
70. Gray, *George Eliot*, 56–57.
71. Weliver, *Women Musicians*, 199.
72. Weliver, 201.
73. Ablow, *Marriage*, 76–78.

74. As Manne writes, one of the most pervasive rape myths is that "real rapists will appear on our radars either as devils, decked out with horns and pitchforks, or else as monsters—that is, as creepy and ghoulish creatures," or that they will be "psychopaths . . . ruthless, unfeeling, and sadistic." Manne, *Down Girl*, 199.

75. Ralph Pite, *Thomas Hardy: The Guarded Life* (New Haven, CT: Yale University Press, 2007), 66.

76. C. M. Jackson-Houlston, "Thomas Hardy's Use of Traditional Song," *Nineteenth-Century Literature* 44, no. 3 (December 1989): 301.

77. Catherine Charlwood, "'Habitually Embodied' Memories: The Materiality and Physicality of Music in Hardy's Poetry," *Nineteenth-Century Music Review* 17 (2020): 245–69; Joan Grundy, *Hardy and the Sister Arts* (London: Macmillan, 1979), 135–36.

78. Gustav Holst to Thomas Hardy (August 4, 1927), Thomas Hardy Archives, Dorset Museum, Dorchester, UK.

79. For more on this debate, see Kivy, *Music*.

80. Thomas Hardy, *Literary Notebooks of Thomas Hardy*, 2 vols., ed. Lennart A. Björk (London: Macmillan, 1985), 1:51.

81. Thomas Hardy, *The Collected Letters of Thomas Hardy*, ed. Richard Little Purdy and Michael Millgate (Oxford: Clarendon Press, 1978), 58. For more on Hardy's engagement with contemporary physiology, see Suzanne Keen, *Thomas Hardy's Brains* (Columbus: Ohio State University Press, 2014); Morgan, *Outward*; Pamela Gossin, *Thomas Hardy's Novel Universe: Astronomy, Cosmology, and Gender in the Post-Darwinian World* (Aldershot: Ashgate, 2007).

82. Hardy, *Literary Notebooks*, 1:92, emphasis in original.

83. See Elaine Scarry, "Work and the Body in Hardy and Other Nineteenth-Century Novelists," *Representations* 3 (Summer 1983): 90; John Hughes, *"Ecstatic Sound": Music and Individuality in the Work of Thomas Hardy* (Aldershot: Ashgate, 2001), 15–16; Morgan, *Outward*, 117–20; Keen, *Thomas Hardy's Brains*.

84. Hardy, *Literary Notebooks*, 2:42.

85. Michael Millgate, *Thomas Hardy's Library at Max Gate: Catalogue of an Attempted Reconstruction*, University of Toronto Library, https://hardy.library.utoronto.ca/.

86. Hardy, *Literary Notebooks*, 1:90; Thomas Hardy, *The Return of the Native*, ed. Simon Avery (Toronto: Broadview, 2013), 51.

87. Hardy, *Literary Notebooks*, 1:495, 174.

88. Hardy, 1:174.

89. See Pite, *Thomas Hardy*; Michael Irwin, *Reading Hardy's Landscapes* (New York: St. Martin's, 2000); Simon Gatrell, *Thomas Hardy's Vision of Wessex* (New York: Palgrave Macmillan, 2003); David James, "Hearing Hardy: Soundscapes and the Profitable Reader," *Journal of Narrative Theory* 40, no. 2 (Summer 2010): 131–55; Hughes, *"Ecstatic Sound,"* 2–5; Clapp-Itynre, *Angelic Airs*, 174, 180; Mark Asquith,

Thomas Hardy, Metaphysics, and Music (Hampshire, UK: Palgrave Macmillan, 2005), 4–5, 11, 90.

90. See Hughes, *"Ecstatic Sound,"* 9; Asquith, *Metaphysics*, 83.
91. Asquith, *Metaphysics*, 167.
92. For more on this critical history, see Spampinato, "Rereading Rape"; Gurman, "Sex, Consent."
93. Thomas Hardy, *Tess of the D'Urbervilles*, ed. Simon Gatrell (Oxford: Oxford World's Classics, 2008), 82.
94. Thomas Hardy, *Desperate Remedies*, ed. Patricia Ingham (Oxford: Oxford University Press, 2003), 87.
95. Hardy, 87.
96. Hardy, 227.
97. Hardy, 126.
98. Hardy, 126.
99. Hardy, 127.
100. Hardy, 129.
101. Hardy, 129.
102. Hardy, 130.
103. Hardy, 130.
104. Hardy, 130–31.
105. Hardy, 130–31.
106. Hardy, 130–31.
107. Hardy, 131–32.
108. Hardy, 132.
109. Hardy, 132.
110. Hardy, 132.
111. Hardy, 133.
112. Hardy, 135.
113. Hardy, 132.
114. Thomas Hardy, "The Fiddler of the Reels," in *Life's Little Ironies*, ed. Simon Gatrell (Oxford: Oxford University Press, 2008), 138.
115. Hardy, 137–38.
116. Hardy, 138.
117. Hardy, 138–39; Gillett, *Musical Women*, 95–97.
118. Hardy, "Fiddler," 138.
119. Hardy, 139–40.
120. Hardy, 138, 152.
121. Hardy, 139.
122. Hardy, 139.
123. Hardy, 140–41, 148.
124. Hardy, 140.

125. Hardy, 140.
126. Gurney, *Power*, 103.
127. Hardy, "Fiddler," 140.
128. Hardy, 140.
129. Hardy, 141.
130. Hardy, *Tess*, 94; Hardy, "Fiddler," 146.
131. Spampinato, "Rereading Rape," 144.
132. Hardy, "Fiddler," 149.
133. Hardy, 149–50.
134. Hardy, 150.
135. Hardy, 150.
136. Hardy, 151.
137. Hardy, 151.
138. Hardy, 150.
139. Hardy, 151.
140. Hardy, 152.
141. Hardy, 152.
142. Hughes, *"Ecstatic Sound,"* 34.
143. Asquith, *Metaphysics*, 92.
144. George Levine, *Reading Thomas Hardy* (Cambridge: Cambridge University Press, 2017), 7.

Chapter Five

1. *Teleny, or the Reverse of the Medal* (New York: Mondial, 2006), 5.
2. *Teleny*, 5.
3. Neil Bartlett, *Who Was That Man? A Present for Mr. Oscar Wilde* (London: Serpent's Tail, 1988), 83. For more on *Teleny*'s publication history, including its tenuous associations with Wilde and rumors that he had authored the novella, see Collette Colligan, *A Publisher's Paradise: Expatriate Literary Culture in Paris, 1890–1960* (Amherst: University of Massachusetts Press, 2013); Ed Cohen, "Writing Gone Wilde: Homoerotic Desire in the Closet of Representation," *PMLA* 102, no. 5 (October 1987): 801–13; Joyce, *LGBT Victorians*.
4. For discussions of the history of the term *homosexual*, see Benjamin Kahan, *The Book of Minor Perverts* (Chicago: University of Chicago Press, 2019), 16; Elaine Showalter, *Sexual Anarchy: Gender and Culture at the Fin de Siècle* (New York: Viking, 1990), 171.
5. Sam Abel, *Opera in the Flesh: Sexuality in Operatic Performance* (Boulder, CO: Westview Press, 1996); Wayne Koestenbaum, *The Queen's Throat: Opera, Homosexuality, and the Mystery of Desire* (Boston: Da Capo Press, 2001), 43. For more on the term *inversion*, see Cohen, "Writing Gone Wilde," 811; George Chauncey,

Gay New York: Gender, Urban Culture, and the Making of the Gay Male World, 1890–1940 (New York: Basic Books, 1994), 48.

6. David Friedman, *Wilde in America: Oscar Wilde and the Invention of Modern Celebrity* (New York: W. W. Norton, 2014), 36; Deutsch, *British Literature*, 142.

7. Havelock Ellis, *Studies in the Psychology of Sex*, vol. 2, 3rd ed. (Philadelphia: F. A. Davis, 1915), 295. For an extensive discussion of the role of music in fin de siècle sexological writing, see Riddell, *Music*, ch. 1.

8. Ellis, *Studies*, 2:295.

9. Riddell, *Music*, 25–26.

10. Deutsch, *British Literature*, 151, 142.

11. Philip Brett, "Musicality, Essentialism, and the Closet," in Brett and Wood, *Queering the Pitch*, 17; Elizabeth Wood, "Sapphonics," in Brett and Wood, *Queering the Pitch*, 27–28.

12. Law, "'Perniciously,'" 196.

13. Wilde, *Dorian Gray*, 140; Sutton, "Music and Sexuality," 214–15.

14. Jarvis, *Exquisite Masochism*, vii.

15. For more on the audiences for pornography in Victorian England, see Steven Marcus, *The Other Victorians: A Study of Sexuality and Pornography in Mid-Nineteenth-Century England* (New York: Basic Books, 1964); Peter Mendes, *Clandestine Erotic Fiction in English, 1800–1930: A Bibliographical Study* (New York: Routledge, 1993); Collette Colligan, *The Traffic in Obscenity from Byron to Beardsley: Sexuality and Exoticism in Nineteenth-Century Print Culture* (London: Palgrave Macmillan, 2006); Colligan, *Publisher's*; Deborah Lutz, "The Secret Rooms of *My Secret Life*," *English Studies in Canada* 31, no. 1 (March 2005), 118–19. For a discussion of pornography's messy archive—and an important corrective to the myth that eighteenth- and nineteenth-century pornography's focus was solely on "heteropenetration," see Kathleen Lubey, *What Pornography Knows: Sex and Social Protest Since the Eighteenth Century* (Stanford, CA: Stanford University Press, 2022), 2, 5.

16. Law acknowledges that *Teleny* portrays same-sex desire more frankly than other texts of the period: "*Teleny* must be set apart from other fiction of the time considered here. No book to be sold above the counter would treat any form of sexuality so explicitly, of course, yet this lurid treatment, written for (and by) a specialized readership, provides a rare undisguised look at the perceived connection between music and homosexuality." Law, "'Perniciously,'" 185. Building on Law's discussion, this chapter investigates how the science of music physics and physiology in particular informs this "lurid treatment."

17. Lynda Nead, *Victorian Babylon: People, Streets and Images in Nineteenth-Century London* (New Haven, CT: Yale University Press, 2005), 84, 109.

18. *The Sins of the Cities of the Plain*, ed. Wolfram Setz (Richmond, VA: Valancourt Classics, 2012), 50.

19. Aubrey Beardsley, *The Story of Venus and Tannhäuser: A Romantic Novel* (For Private Circulation, 1907), 86.

20. *The Adventures of a School-Boy*, in *The New Epicurean and The Adventures of a School-Boy: Two Tales from the Victorian Underground* (New York: Grove Press, 1984), 123.

21. Deutsch, *British Literature*, 147; Riddell, *Music*, 18.

22. Joseph Bristow, "'A Few Drops of Thick, White, Viscid Sperm': *Teleny* and the Defense of the Phallus," in *Porn Archives*, ed. Tim Dean et al. (Durham. NC: Duke University Press, 2014), 155.

23. For further readings of *Teleny* as a political work, see Cohen, "Writing Gone Wilde"; Bristow, "Drops."

24. See Caleb, introduction to *Teleny*. The authors of *Teleny* invoke this language when Des Grieux asks, "Had I committed a crime against nature when my own nature found peace and happiness thereby?" *Teleny*, 86.

25. See Riddell, *Music*, 25–26.

26. Taylor, "Sound Desires," 277.

27. Cohen, "Writing Gone Wilde," 804; Bristow, "Drops," 145–46.

28. *Teleny*, 33; Kahan, *Perverts*, 23. For more on congenitality in *Teleny*, see Joyce, *LGBT Victorians*, 235.

29. *Teleny*, 19, 23, 34.

30. See Michael Warner, *The Trouble with Normal: Sex, Politics, and the Ethics of Queer Life* (Cambridge, MA: Harvard University Press, 1999), 9; Friedman, *Before Queer Theory*, 15.

31. See Stephen Jay Gould, *The Mismeasure of Man* (New York: W. W. Norton, 1996); James Poskett, *Materials of the Mind: Phrenology, Race, and the Global History of Science, 1815–1920* (Chicago: University of Chicago Press, 2019).

32. Friedman, *Before Queer Theory*, 14.

33. Diane Mason, *The Secret Vice: Masturbation in Victorian Fiction and Medical Culture* (Oxford: Oxford University Press, 2006), 9; Bristow, "Drops," 157–58; Matt Cook, *London and the Culture of Homosexuality, 1885–1914* (Cambridge: Cambridge University Press, 2003), 105; Christopher Wellings, "Dangerous Desires: The Uses of Women in *Teleny*," *Oscholars: Special* Teleny *Issue* (Autumn 2008), http://www.oscholars.com/Teleny/wellings.htm.

34. Ahmed, *What's the Use?*, 199.

35. See Wilson, *Gut Feminism*; Roy, "Somatic Matters"; Willey, *Undoing Monogamy*.

36. *Teleny*, 8.

37. Colligan, *Publisher's*, 237–38; Bristow, "Drops," 155–57; Pamela Thurschwell, *Literature, Technology, and Magical Thinking, 1880–1920* (Cambridge: Cambridge University Press, 2001), 34–35; Ann Gagné, *Embodying the Tactile in Victorian Literature: Touching Bodies/Bodies Touching* (Lanham, MD: Rowman and Littlefield, 2021), 86.

38. *Teleny*, 5.

39. Riddell, *Music*, 154; Colligan, *Publisher's*, 231–35; Benjamin Bagocius, "Masturbation and Physiological Romance in *Teleny*," *Criticism* 59, no. 3 (2017): 441.
40. Gurney, *Power*, 103, 167–68.
41. *Teleny*, 2–3.
42. *Teleny*, 4.
43. See Bristow, "Drops"; Riddell, *Music*.
44. Helmholtz, *Sensations*, 129.
45. For a theorization of the earworm as it relates to listening technologies, the attention economy, and capitalism, see Eldritch Priest, *Earworm and Event* (Durham, NC: Duke University Press, 2022).
46. *Teleny*, 18.
47. Gurney, *Power*, 155.
48. Gurney, *Power*, 375.
49. Sacks, *Musicophilia*, 56.
50. *Teleny*, 92.
51. *Teleny*, 13.
52. *Teleny*, 92.
53. *Teleny*, 93.
54. See Fretwell, *Sensory Experiments*; Zon, *Evolution*; Hui, *Psychophysical Ear*.
55. Ellis, *Studies*, 4:122; Chomet, *Influence*, 190.
56. *Teleny*, 2.
57. *Teleny*, 5.
58. Colligan, *Publisher's*, 213.
59. *Teleny: Étude Physiologique*, trans. Charles Hirsch (Paris: La Musardine, 2009), 27. Translation mine.
60. *Teleny*, 5.
61. *Teleny*, 62.
62. *Teleny*, 33.
63. Cusick, "Lesbian," 78–79, emphasis in original. For a discussion of the 1991 conference, see Maus, "'What If Music IS Sex?'"
64. Cusick, "Lesbian," 70; Taylor, "Taking It in the Ear," 603.
65. Taylor, "Taking It in the Ear," 609.
66. Taylor, "Sound Desires," 277.
67. *Teleny*, 4–5.
68. *Teleny*, 4.
69. *Teleny: Étude physiologique*, 27–28. "The notes murmured in my ears, with the panting of a feverish lust, the noise of a run of kisses. My entire body convulsed into an erotic rage. I had dry lips, panting breath, stiff limbs, swollen veins; and yet I remained impassive like those around me. . . . Dizziness seized my brain; a burning lava coursed through my veins, a few drops spurted out. . . . I palpitated. . . . I started, I was trembling." Translation mine.

70. "Baiser," *Le Petit Robert de la Langue Française*, accessed June 13, 2019, https://dictionnaire.lerobert.com/definition/baiser.

71. *Teleny*, 15.

72. Cusick, "Lesbian," 70.

73. David Halperin, *Saint Foucault: Towards a Gay Hagiography* (Oxford: Oxford University Press, 1995), 88.

74. Dean, *Beyond Sexuality*, 277.

Chapter Six

1. Charles Dickens, *Dombey and Son*, edited by Andrew Sanders (London: Penguin Classics, 2002), 194, 886.

2. Dickens, 194, 886.

3. Dickens, 881.

4. Dickens, 882.

5. Dickens, 881, 886.

6. Dickens, 886. Here, Dickens refers to the final movement of George Friedrich Handel's Suite No. 5 in E Major (1720), a piece that would later play a role in *Great Expectations* (1860–61) when Herbert Pocket nicknames Pip "Handel" due to his training as a blacksmith.

7. Jeremy Chow, "Mellifluent Sexuality: Female Intimacy in Ann Radcliffe's *The Romance of the Forest*," *Eighteenth-Century Fiction* 30, no. 2 (Winter 2017–18): 215.

8. Riddell, *Music*, 139.

9. Thomas, *Violin-Player*, III.320.

10. Herman Melville, *Pierre: Or, The Ambiguities* (London: Penguin Books, 1996), 125; Qtd. in Gillett, *Musical Women*, 93.

11. Marion Scott, *Violin Verses* (London: Walter Scott, 1905), 6.

12. Dickens, *Dombey*, 194, 886.

13. Dickens, 881, 886.

14. As Anna Henchman argues, Victorian writers such as Hardy and Hopkins developed strikingly "porous" notions of "sentience," which they ascribed not just to living, human beings but also to dead bodies, plants, tissue, and other matter, creating epistemological confusion between beings and nonbeings. Anna Henchman, "Sentience," *Victorian Literature and Culture* 46, nos. 3–4 (2018): 861–65.

15. Jane Bennett, *Vibrant Matter: A Political Ecology of Things* (Durham, NC: Duke University Press, 2010), 6.

16. Elaine Freedgood, *The Ideas in Things: Fugitive Meaning in the Victorian Novel* (Chicago: University of Chicago Press, 2006), 5. For other work on the agency of Victorian "things," see Katharina Boehm, "Introduction: Bodies and Things," in *Bodies and Things in Nineteenth-Century Literature and Culture*, ed. Katharina Boehm

(London: Palgrave Macmillan, 2012), 1–16; Isobel Armstrong, "Bodily Things and Thingly Bodies: Circumventing the Subject-Object Binary," in Boehm, *Bodies and Things*, 17–44; Cohen, *Embodied*; Lyn Pykett, "The Material Turn in Victorian Studies," *Literature Compass* 1, no. 1 (2004): 1–5; Catherine Gallagher and Stephen Greenblatt, *Practicing New Historicism* (Chicago: University of Chicago Press, 2001).

17. Bennett, *Vibrant Matter*, xvi. The depictions of instruments in these texts also resonate with Bruno Latour's famous assertion that "objects, too, have agency." Bruno Latour, *Reassembling the Social: An Introduction to Actor-Network Theory* (Oxford: Oxford University Press, 2005), 63.

18. Rosi Braidotti, *Posthuman Feminism* (Cambridge: Polity, 2022), 6, 112, 134–35. It is crucial to acknowledge feminist and queer posthumanism's indebtedness to Indigenous and anticolonial scholarship that has long explored the interrelations between humans and the environment. See Kim Tallbear, "An Indigenous Reflection on Working Beyond the Human/Not Human," *GLQ* 21, nos. 2–3 (June 2015): 230–35. See also Jinthana Haritaworn, "Decolonizing the Non/Human," *GLQ* 21, nos. 2–3 (June 2015): 210–13.

19. Patricia MacCormack, "Queer Posthumanism: Cyborgs, Animals, Monsters, Perverts," in *The Ashgate Research Companion to Queer Theory*, ed. Noreen Giffney and Michael O'Rourke (Aldershot: Ashgate, 2009), 112.

20. Riddell, *Music*, 139.

21. Freeman, *Time Binds*, 53. See also Lynne Huffer, *Are the Lips a Grave? A Queer Feminist on the Ethics of Sex* (New York: Columbia University Press, 2013); Willey, *Undoing Monogamy*; Przybylo, *Asexual Erotics*; Dean, *Beyond Sexuality*.

22. Eileen Joy, "Improbable Manners of Being," *GLQ* 21, nos. 2–3 (June 2015): 224.

23. Katherine Behar, "An Introduction to OOF," in *Object-Oriented Feminism*, ed. Katherine Behar (Minneapolis: University of Minnesota Press, 2016), 16.

24. Cusick, "Lesbian," 78; Pedro Rebelo, "Haptic Sensation and Instrumental Transgression," *Contemporary Music Review* 25, no. 1/2 (February/April 2006): 31–32.

25. See Judith Butler, "Is Kinship Always Already Heterosexual?," *differences* 13, no. 1 (2002): 14–44; Stefani Engelstein, *Sibling Action: The Genealogical Structure of Modernity* (New York: Columbia University Press, 2007); Sarah Franklin and Susan McKinnon, introduction to *Relative Values: Reconfiguring Kinship Studies*, ed. Sarah Franklin and Susan McKinnon (Durham, NC: Duke University Press, 2002): 1–28; Tyler Bradway and Elizabeth Freeman, "Introduction: Kincoherence/Kin-aesthetics/Kinematics," in *Queer Kinship: Race, Sex, Belonging, Form*, ed. Tyler Bradway and Elizabeth Freeman (Durham, NC: Duke University Press, 2022), 1–24; Mahoney, *Queer Kinship*, 8–14; Ahmed, *Queer Phenomenology*, 107.

26. Paul Sanden, "Hearing Glenn Gould's Body: Corporeal Liveness in Recorded Music," *Current Musicology* 88 (Fall 2009): 25.

27. Sanden, 25.

28. Rebelo, "Haptic," 27, 32; Sile O'Modhrain and R. Brent Gillespie, "Once More, with Feeling: Revisiting the Role of Touch in Performer-Instrument Interaction," in *Musical Haptics*, ed. Stefano Papetti and Charalampos Saitis (Berlin: Springer, 2018), 11–12. I am grateful to Catherine Charlwood's recent article in *Nineteenth-Century Music Review* for alerting me to the work of O'Modhrain and Gillespie and contemporary research on musical haptics. Charlwood, "'Habitually Embodied.'" See also Cook, *Beyond the Score*, 314–15.

29. Atau Tanaka and Marco Donnarumma, "The Body as Musical Instrument," in Kim and Gilman, *Oxford Handbook of Music and the Body*, 79; Rebelo, "Haptic," 30; O'Modhrain and Gillespie, "Once More, with Feeling," 11–15.

30. Maria M. Delgado, "Carles Santos, 'Music in the Theatre,'" in *Taking It to the Bridge: Music as Performance*, ed. Nicholas Cook and Richard Pettengill (Ann Arbor: University of Michigan Press, 2013), 243–44.

31. Rebelo, "Haptic," 30.

32. Ahmed, *Queer Phenomenology*, 3, 23.

33. James Q. Davies, *Romantic Anatomies of Performance* (Berkeley: University of California Press, 2014); Dana Gooley, "The Battle Against Instrumental Virtuosity in the Early Nineteenth Century," in *Franz Liszt and His World*, ed. Christopher H. Gibbs and Dana Gooley (Princeton, NJ: Princeton University Press, 2006), 75–112; Ross, "Hold Your Applause."

34. Helmholtz, *Sensations*, 8.

35. Xaver Scharwenka, "The Octave Stacatto," in *The Music of the Modern World, Illustrated in the Lives and Works of the Greatest Modern Musicians and in Reproduction of Famous Paintings, Etc.*, ed. Anton Seidl et al. (Boston: D. Appleton, 1895), 5, emphasis in original.

36. "On the Physiology of Pianoforte Playing, with a Practical Application of a New Theory," *Musical Standard* 32, no. 14 (January 1888): 18–19; E. S. J. van der Straeten, "The Technics of Violoncello Playing," *The Strad* 6, no. 61 (May 1895): 19; Scharwenka, "Octave," 5.

37. van der Straeten, "Technics," 19.

38. Jackson, *Harmonious Triads*, 234–35.

39. "Correction of Faulty Touch," *Musical Herald*, no. 498 (April 1889): 82.

40. "The Organ World: On the Physiology of Pianoforte Playing, with a Practical Application of a New Theory," *Musical World* 68, no. 3 (November 3, 1888): 849.

41. Advertisement for the dactylergon, *English Mechanic and World of Science*, no. 1295 (January 17, 1890): 432.

42. Jackson, *Harmonious Triads*, 235; Advertisement for the technicon, *Monthly Musical Record* (1 May 1887): 119.

43. Davies, *Romantic Anatomies*, 113.

44. "Dr. W. H. Stone on the Causes of Rise in Orchestral Pitch and the Methods of Obviating It," *Monthly Musical Record* 11 (May 1, 1881): 90.

45. "The Importance of the Nose in Connexion with the Practice of Singing, and Playing Wind Instruments," *Musical World* 32, no. 41 (October 14, 1854): 685, emphasis in original.

46. Gillett, *Musical Women*, 87; Christina Bashford, "Sensuality and the Senses: The Appeal of the Violin in Victorian Culture" (conference paper presented at the Midwest Victorian Studies Association, Iowa City, Iowa, May 1, 2015); "Nerves of the Violin," *The Strad* 5, no. 57 (January 1895): 279.

47. "Nerves," 279.

48. "Nerves," 279. This article also included the sexist comment that violins are "so uncertain and so capricious" that "it is a matter for wonder that they have not long since been classed, like ships, in the feminine gender."

49. George Eliot, "Stradivarius," in *The Legend of Jubal and Other Poems* (Boston: James R. Osgood, 1874), 212; George Eliot, *Middlemarch*, ed. Rosemary Ashton (London: Penguin Books, 1994), 838.

50. Eliot, "Stradivarius," 212.

51. Eliot, 207.

52. Eliot, *Mill*, 389.

53. Eliot, 390.

54. Eliot, 417–18.

55. Da Sousa Correa, *George Eliot*, 111.

56. Lewes, *Physiology*, 56.

57. Weliver, *Women Musicians*, 197.

58. Ahmed, *Queer Phenomenology*, 23.

59. Da Sousa Correa, *George Eliot*, 103.

60. Anthony Trollope, *The Letters of Anthony Trollope, 1851–1882*, ed. Bradford Allen Booth (Oxford: Oxford University Press, 1951).

61. Anthony Trollope, *The Warden*, ed. Nicholas Shrimpton (Oxford: Oxford University Press, 2009), 81.

62. Trollope, 9.

63. Trollope, 20.

64. Trollope, 81.

65. Trollope, 20, 22, 50–51.

66. Trollope, 51.

67. Trollope, 39.

68. Zon, *Evolution*, 302–3; Jarlath Killeen, "Emptying Time in Anthony Trollope's *The Warden*," in *Victorian Time: Technologies, Standardization, Catastrophes*, ed. Trish Ferguson (London: Palgrave Macmillan, 2013), 51.

69. Killeen, "Emptying Time," 51.

70. Bo Earle, "Policing and Performing Liberal Individuality in Anthony Trollope's *The Warden*," *Nineteenth-Century Literature* 61, no. 1 (June 2006): 28.

71. Lewes, *Physiology*, 56; Spencer, *Principles*, 451.

72. Trollope, *Warden*, 142.

73. Zon, *Evolution*, 303; Earle, "Policing," 7.

74. Thomas Hardy, *Under the Greenwood Tree*, ed. Claire Seymour (London: Wordsworth, 1994), 33–34, 36–37; Thomas Hardy, *Far from the Madding Crowd*, ed. Rosemary Morgan and Shannon Russell (London: Penguin, 2003), 210 (hereafter cited as *FFTMC*).

75. Hardy, *FFTMC*, 211.

76. Charlwood, "'Habitually Embodied,'" 255–56; Elaine Auyoung, "Phantoms and Fictional Persons: Hardy's Phenomenology of Loss," *Victorian Studies* 59, no. 3 (Spring 2017): 401. Hardy was not the first poet to write from the perspective of musical instruments. Scott's *Violin Verses* includes several poems written "by" Guadagnini, Amati, Nicholas Gagliano, and Guarnerius violins, all of whom have their own personalities. The last few pages of *Violin Verses* contain a collection of "Fiddle Philosophies," in which the instruments expound upon various moral and intellectual theories. Scott, *Violin Verses*, 14, 16–19, 29–36.

77. See Tim Armstrong, "Hardy, History, and Recorded Music," in *Thomas Hardy and Contemporary Literary Studies*, ed. Tim Dolin and Peter Widdowson (London: Palgrave Macmillan, 2004), 161, 166; Charlwood, "'Habitually Embodied,'" 259–62; Auyoung, "Phantoms," 401; Riddell, *Music*, 151.

78. Auyoung, "Phantoms," 401.

79. Thomas Hardy, "Haunting Fingers," in *Thomas Hardy: The Complete Poems*, ed. James Gibson (London: Palgrave Macmillan, 2001), 590–92.

80. See Charlwood, "'Habitually Embodied.'"

81. For queer readings of Hardy, see Richard Dellamora, "Male Relations in Thomas Hardy's *Jude the Obscure*," *Papers on Language and Literature* 27, no. 4 (Fall 1991): 453–72; Brett Neilson, "Hardy, Barbarism, and the Transformations of Modernity," in Dolin and Widdowson, *Thomas Hardy and Contemporary Literary Studies*, 65–80; Kate Thomas, *Postal Pleasures* (Oxford: Oxford University Press, 2012). For posthuman readings of Hardy, see Cohen, *Embodied*; Anna West, *Thomas Hardy and Animals* (Cambridge: Cambridge University Press, 2017).

82. Hardy, "Haunting Fingers," 591.

83. Hardy, 590.

84. Hardy, 591.

85. Hardy, 590.

86. Cox, *Touch*, 171.

87. Hardy, "Haunting Fingers," 590.

88. Hardy, 591. See Charlwood, "'Habitually Embodied'"; and Armstrong, "Hardy, History."

89. Ahmed, *Queer Phenomenology*, 54.

90. Levine, *Reading*, 8.

91. Latour, *Reassembling*, 63. For discussions of music technology in modernist literature, see Brad Bucknell, *Literary Modernism and Musical Aesthetics: Pater, Pound, Joyce, and Stein* (Cambridge: Cambridge University Press, 2001); Josh Epstein,

Sublime Noise: Musical Culture and the Modernist Writer (Baltimore: Johns Hopkins University Press, 2014).

Chapter Seven

1. John Meade Falkner, *The Lost Stradivarius* (New York: D. Appleton, 1896), 141.

2. Richard Marsh, "The Violin," in *The Seen and the Unseen* (London: Methuen, 1900), 92–115. For more on this text, see Riddell, *Music*, 146–47.

3. E. M. Forster, "Dr. Woolacott," in *The Life to Come and Other Stories* (New York: Norton, 1972), 93. For more on this scene, see Riddell, *Music*, 154–57.

4. David Toop, *Sinister Resonance: The Mediumship of the Listener* (London: Bloomsbury, 2010), xv.

5. Michael Cox and R. A. Gilbert, introduction to *The Oxford Book of Victorian Ghost Stories*, ed. Michael Cox and R. A. Gilbert (Oxford: Oxford University Press, 2003), xi–xii; Eve Lynch, "Spectral Politics: The Victorian Ghost Story and the Domestic Servant," in *The Victorian Supernatural*, ed. Nicola Bown, Carolyn Burdett, and Pamela Thurschwell (Cambridge: Cambridge University Press, 2004), 68.

6. Cox and Gilbert, introduction, xii; Lynch, "Spectral Politics," 68.

7. Julian Wolfreys, *Victorian Hauntings: Spectrality, Gothic, the Uncanny and Literature* (London: Palgrave Macmillan, 2002), 1, 7; Cox and Gilbert, introduction, x.

8. Wolfreys, *Victorian Hauntings*, 7.

9. Dinah Mulock, "The Last House in C—— Street," in Cox and Gilbert, *Oxford Book of Victorian Ghost Stories*, 44; Bernard Capes, "An Eddy on the Floor," in Cox and Gilbert, *Oxford Book of Victorian Ghost Stories*, 408; Rhoda Broughton, "The Truth, the Whole Truth, and Nothing but the Truth," in Cox and Gilbert, *Oxford Book of Victorian Ghost Stories*, 78.

10. Peter Buse and Andrew Stott, "Introduction: A Future for Haunting," in *Ghosts: Deconstruction, Psychoanalysis, History*, ed. Peter Buse and Andrew Stott (London: Palgrave Macmillan, 1999), 1.

11. Wolfreys, *Victorian Hauntings*, 24.

12. Wolfreys, 24.

13. Roger Freitas, "The Eroticism of Emasculation: Confronting the Baroque Body of the Castrato," *Journal of Musicology* 20, no. 2 (Spring 2003): 214. Similarly, Emily Wilbourne notes that the castrato is "quintessentially queer, for he slips between the cracks of heteronormativity: he modifies his body to suit his desire and he abdicates the procreative responsibilities of the heterosexual. At the heart of this identity is a phenomenal, superhuman voice." Emily Wilbourne, "The Queer History of the Castrato," in Maus and Whiteley, *Oxford Handbook of Music and Queerness*, 448–49.

14. For a reading of the violin in *The Lost Stradivarius* as possessing its own agency, see Eliot Bates, "The Social Life of Musical Instruments," *Ethnomusicology* 56, no. 3 (Fall 2012): 363–95.

15. Vernon Lee, "A Wicked Voice," in *Hauntings and Other Fantastic Tales*, ed. Catherine Maxwell and Patricia Pulham (Peterborough, Ontario: Broadview, 2006), 179, 174.

16. Terry Castle, *The Apparitional Lesbian: Female Homosexuality and Modern Culture* (New York: Columbia University Press, 1993), 1; Diana Fuss, *Dying Modern: A Meditation on Elegy* (Durham, NC: Duke University Press, 2013), 2–3; Muñoz, *Cruising Utopia*, 46. See also Jack Halberstam, *Skin Shows* (Durham, NC: Duke University Press, 1995), 3; Carla Freccero, *Queer/Early/Modern* (Durham, NC: Duke University Press, 2006), 85.

17. Lee, "Wicked Voice," 179.

18. Freccero, *Queer/Early/Modern*, 85.

19. Lee, "Wicked Voice," 168, 171, 156.

20. Lee Edelman, *No Future: Queer Theory and the Death Drive* (Durham, NC: Duke University Press, 2004), 2–4; Freeman, *Time Binds*, 10.

21. Freeman, *Time Binds*, 104.

22. Freeman, 105.

23. Carolyn Dinshaw, *Getting Medieval: Sexualities and Communities, Pre- and Postmodern* (Durham, NC: Duke University Press, 1999), 3.

24. Stephen Best and Sharon Marcus, "Surface Reading: An Introduction," *Representations* 108, no. 1 (Fall 2009): 3.

25. Sylvia Miezkowski, *Resonant Alterities: Sound, Desire and Anxiety in Non-Realist Fiction* (New York: Columbia University Press, 2014), 45–46; Ruth Bienstock Anolik, "Sexual Horror: Fears of the Sexual Other," in *Horrifying Sex: Essays on Sexual Deviance in Gothic Literature*, ed. Ruth Bienstock Anolik (Jefferson, NC: McFarland, 2007), 7; Cox and Gilbert, introduction, ix.

26. Edmund Craster, "Personal Note on John Meade Falkner," in *The Nebuly Coat and The Lost Stradivarius* by John Meade Falkner (Oxford: Oxford University Press, 1954), xi–xii.

27. Nicholas Daly, "Somewhere There's Music: John Meade Falkner's *The Lost Stradivarius*," in *Romancing Decay: Ideas of Decadence in European Culture*, ed. Michael St. John (New York: Routledge, 2016), ch. 7, Google Books.

28. G. M. Young, introduction to *The Nebuly Coat and The Lost Stradivarius*, by John Meade Falkner (Oxford: Oxford University Press, 1954), vii.

29. Falkner, *Lost Stradivarius*, 61.

30. Falkner, 61.

31. Falkner, 3–4.

32. Falkner, 90.

33. Falkner, 93.

34. See Riddell, *Music*, ch. 1.

35. Falkner, *Lost Stradivarius*, 1, 200.
36. Falkner, 224.
37. Falkner, 212, 69.
38. Falkner, 116, 128.
39. Falkner, 119–20.
40. Falkner, 135.
41. Falkner, 34, 36. Haweis saw music as a "vast civiliser, recreator, health-giver, work-inspirer, and purifier of man's life." H. R. Haweis, *My Musical Life*, 8th ed. (New York: Longmans, Green, 1912), 194.
42. Daly, "Somewhere There's Music."
43. Indeed, as Daly points out, "we don't actually have Sir John Maltravers' version of events, nor do we have a 'neutral' third-person narrator"; for Daly, reading Sophia and William as "unreliable narrators" opens up the possibility that the novel might actually offer a "veiled but affirmative account of male love." Daly, "Somewhere There's Music."
44. Falkner, *Lost Stradivarius*, 6.
45. Falkner, 14.
46. Though most of Helmholtz's theories emerged after the time in which *The Lost Stradivarius* is set (1841 and the years immediately following), there would have been a robust culture of acoustics at Oxford by the time Falkner was writing the novel. Rosemary Golding, *Music and Academia in Victorian Britain* (London: Routledge, 2016), 5; Hiebert, "Listening to the Piano Pedal," 240. Falkner, *Lost Stradivarius*, 7.
47. Falkner, 10–11.
48. Falkner, 13.
49. Falkner, 15.
50. Falkner, 15.
51. Falkner, 60, 57.
52. Falkner, 53.
53. Helmholtz, *Sensations*, 23–24.
54. Helmholtz, 22.
55. Tyndall, *Sound*, 3rd ed., 403.
56. Falkner, *Lost Stradivarius*, 104.
57. Falkner, 104–5.
58. Falkner, 26.
59. Falkner, 108.
60. Falkner, 138.
61. Falkner, 138.
62. Falkner, 139.
63. Falkner, 139–41.
64. Falkner, 179.
65. Falkner, 239.

66. Lee, "Wicked Voice," 161.
67. Lee, 181.
68. Lee, 155.
69. See Catherine Maxwell, "Sappho, Mary Wakefield, and Vernon Lee's 'A Wicked Voice,'" *Modern Language Review* 102, no. 4 (October 2007): 960–74; Carlo Caballero, "'A Wicked Voice': On Vernon Lee, Wagner, and the Effects of Music," *Victorian Studies* 35, no. 4 (Summer 1992): 385–408; Angela Leighton, "Ghosts, Aestheticism, and 'Vernon Lee,'" *Victorian Literature and Culture* 28, no. 1 (2000): 1–14; Patricia Pulham, *Art and the Transitional Object in Vernon Lee's Supernatural Tales* (Aldershot: Ashgate, 2008), 17, 23–25.
70. Maxwell, "Sappho," 960.
71. Miezkowski, *Resonant Alterities*, 45–46, 64.
72. Vineta Colby, *Vernon Lee: A Literary Biography* (Charlottesville: University of Virginia Press, 2003); Caballero, "'Wicked Voice,'" 386.
73. Leighton, "Ghosts," 5. For more on scientific aesthetics in Lee's work, see Shafquat Towheed, "The Science of Musical Memory: Vernon Lee and the Remembrance of Sounds Past," in *Words and Notes in the Long Nineteenth Century*, ed. Phyllis Weliver and Katharine Ellis (Suffolk, UK: Boydell and Brewer, 2013), 104–22.
74. Lee, *Music and Its Lovers*, 152.
75. Caballero, "'Wicked Voice,'" 389. Also relevant is Davies's point that castrati were often associated with ghostliness or otherworldliness—a "phantasmic charge . . . the castrato becoming unnerving enough to provoke apprehensions of nonbeing." Davies, *Romantic Anatomies*, 14–15. Of course, in "A Wicked Voice," the castrato *is* a ghost—a fact that further establishes the queerness of the musical encounters.
76. Lee, "Wicked Voice," 166–67.
77. Lee, 168.
78. Lee, 168–69.
79. Lee, 170.
80. See Riddell, *Music*, ch. 1.
81. Lee, "Wicked Voice," 170.
82. Lee, 170.
83. Lee, 173–74.
84. See Pulham, *Art*, 16.
85. Lee, "Wicked Voice," 177.
86. Lee, 177.
87. Lee, 179.
88. Lee, 179.
89. Lee, 179.
90. Lee, 179–80. Vernon Lee first devised a version of this moment in her short story "Winthrop's Adventure" (1881), in which the protagonist hears (and

eventually sees) a harpsichord-playing ghost. Vernon Lee, "Winthrop's Adventure," in *A Phantom Lover and Other Dark Tales*, ed. Mike Ashley (London: British Library, 2020), Kindle.

91. Judith Butler, *Undoing Gender* (New York: Routledge, 2004), 3–4.
92. Lee, "Wicked Voice," 179.
93. Le Guin, *Boccherini's Body*, 24.

Coda

1. *Saturday Night Live*, "Orchestra—SNL," YouTube video, April 16, 2022, 4:43, https://www.youtube.com/watch?v=KhctLo_qS10.
2. Garrett McQueen, "SNL's Public Indictment of Orchestra Culture," *Represent Classical*, April 23, 2022, https://representclassical.com/2022/04/23/snl-public-indictment-of-orchestra-culture/.
3. Levitin, *This Is Your Brain*.
4. Ross, "Hold Your Applause"; Johnson, *Who Needs Classical Music?*
5. Ross, "Hold Your Applause."
6. E. M. Trevenen Dawson, "How to Behave at Concerts," *Monthly Musical Record* 28, no. 326 (February 1, 1898): 28.
7. Qtd. in Levine, *Highbrow/Lowbrow*, 189.
8. Nick Clark, "Cacophony of Sound? People Cough on Purpose During Classical Concerts, Research Finds," *The Independent*, January 29, 2013, https://www.independent.co.uk/arts-entertainment/classical/news/cacophony-sound-people-cough-purpose-during-classical-concerts-research-finds-8471735.html.
9. Martin Kettle, "Prom 71: BBCSO/Vänskä—Review," September 5, 2013, https://www.theguardian.com/music/2013/sep/05/prom-71-bbcso-vanska-review.
10. For more on the histories of concert etiquette, see Levine, *Highbrow/Lowbrow*; Johnson, *Listening in Paris*. See also Draucker, "Ladies' Orchestras."
11. James Bennett II, "What's Up with All Those Ricola Cough Drops at Carnegie Hall?" *WQXR: How to Classical*, June 6, 2018, https://www.wqxr.org/story/ricola-cough-drops-carnegie-hall/.
12. Anne Midgette, "How (Not) to Behave: Manners and the Classical Music Audience," *Washington Post*, April 29, 2015, https://www.washingtonpost.com/entertainment/music/how-not-to-behave-manners-and-the-classical-music-audience/2015/04/29/111a2aa4-bc72-11e4-b274-e5209a3bc9a9_story.html.
13. Robin Bickerstaff Glover, "Symphony Etiquette," Maryland Symphony Orchestra, https://www.marylandsymphony.org/plan-your-visit/what-expect.
14. Fiona Maddocks and Sasha Valeri Millwood, "Should You Go to a Concert with a Cough?," *The Guardian*, December 6, 2014, https://www.theguardian.com/commentisfree/2014/dec/06/should-you-go-to-concert-cough-kyung-wha-chung.

15. Maddocks and Millwood, "Concert with a Cough?"

16. Mary Katherine Wilson, Sarah Marczynski, and Elizabeth O'Brien, "Ethical Behavior of the Classical Music Audience," *Ethical Human Psychology and Psychiatry* 16, no. 2 (2014): 121.

17. Ross, "Hold Your Applause."

18. Cheng uses the term "anonymous" rather than "blind" to avoid the ableist metaphor and to emphasize "the concealment of the performer's identity rather than the direct obstruction of juries' sensory faculties." Cheng, *Loving*, 64. Gene Chieffo, "Blind Symphonic Auditions: Mitigating the Body's Effect on Orchestral Jobs," *Bodylore* (blog), March 4, 2020, https://sites.wp.odu.edu/bodylore/2020/03/04/blind-symphonic-auditions-mitigating-the-bodys-effect-on-orchestral-jobs/.

19. Cheng, *Loving*, 65.

20. Qtd. in Cheng, 64.

21. Maddocks and Millwood, "Concert with a Cough?"

22. TreasuresDelightsEtc, "Music Teacher Notebook," Etsy, accessed September 9, 2023, https://www.etsy.com/sg-en/listing/211797339/notebook-music-is-my-escape-music.

23. Cheng, *Just Vibrations*, 33.

24. Anthony Tommasini, "To Make Orchestras More Diverse, End Blind Auditions," *New York Times*, July 16, 2020, https://www.nytimes.com/2020/07/16/arts/music/blind-auditions-orchestras-race.html.

25. Tommasini, "Make Orchestras More Diverse."

26. Cheng, *Loving*, 66–68.

27. Cheng, 66.

28. Jay Nordlinger, "The Yuja 'n' Jaap Show," *New Criterion* (blog), March 5, 2018, https://newcriterion.com/blogs/dispatch/the-yuja-n-jaap-show.

29. Alex Hawgood, "Hahn-Bin Straddles Classical Music and Fashion," *New York Times*, February 23, 2011, https://www.nytimes.com/2011/02/24/fashion/24HAHNBIN.html.

30. Jeanette Der Bedrosian, "Rock On—Just Not Too Hard," *Johns Hopkins Magazine*, Summer 2019, https://hub.jhu.edu/magazine/2019/summer/guitar-injury-smartguitar-serap-bastepe-gray/.

31. Cheng, *Loving*, 94.

32. For more, see Shannon Draucker, "Food for the Soul, Art of the Flesh: Classical Music, COVID-19, and the Body," *BLARB* (blog), *Los Angeles Review of Books*, September 2, 2021, https://blog.lareviewofbooks.org/essays/food-soul-art-flesh-classical-music-covid-19-body/.

33. L. Hamner et al., "High SARS-CoV-2 Attack Rate Following Exposure at a Choir Practice—Skagit County, Washington, March 2020," *Morbidity and Mortality Weekly Report* 69, no. 19 (May 15, 2020): 606–10; Emily Anthes, "Musical Chairs? Swapping Seats Could Reduce Orchestra Aerosols," *New York Times*, June 23, 2021, https://www.nytimes.com/2021/06/23/health/coronavirus-orchestra.html.

34. "Itzhak Perlman, Violin," University of Georgia Performing Arts Center, April 29, 2023, https://pac.uga.edu/event/itzhak-perlman-violin/; "Itzhak Perlman in Recital," *Cincinnati Symphony Orchestra*, April 10, 2022, https://www.cincinnatisymphony.org/tickets-and-events/buy-tickets/cso/2122-cso-season/itzhak-perlman-in-recital/.

35. Sedgman, *Reasonable Audience*, 6.

36. Lukkas Krohn-Grimberghe, "We Must Breathe—Why It Is Important to Talk About Race and Racism in the Context of Classical Music," *WXQR* (blog), June 2, 2020, https://www.wqxr.org/story/we-must-breathe/.

37. Simpson, "Tics," 227.

38. Simpson, 231.

39. Maddy Shaw Roberts, "The 9-Year-Old Boy Who Blurted Out 'Wow!' at the End of a Concert Has Been Found," *Classic FM*, May 10, 2019, https://www.classicfm.com/music-news/orchestra-seeks-wow-child-concert/.

40. Simpson, "Tics," 233.

41. Helene Gjerris and Bent Nørgaard, "Improving the Performance of Classical Musicians through Interdisciplinary Exercises," *Arts and Humanities as Higher Education*, August 2016, http://www.artsandhumanities.org/improving-the-performance-of-classical-musicians-through-interdisciplinary-exercises/.

42. Gjerris and Nørgaard, "Improving."

43. Astrid Baumgardner, "Engaging Today's Audience: 3 Things That Can Make a Difference to Classical Music," *Astrid Baumgardner* (blog), November 29, 2012, https://www.astridbaumgardner.com/blog-and-resources/blog/engaging-today-u2019s-audience-3-things-that-can-make-a-difference-to-classical-music/.

44. Anthony McGill et al., "Musicians on How to Bring Racial Equity to Auditions," *New York Times*, September 10, 2020, https://www.nytimes.com/2020/09/10/arts/music/diversity-orchestra-auditions.html.

45. McGill et al., "Musicians."

46. Fiona Maddocks, "Yuja Wang: 'If the Music Is Beautiful and Sensual, Why Not Dress to Fit?'" *The Guardian*, April 9, 2017, https://www.theguardian.com/music/2017/apr/09/yuja-wang-piano-interview-fiona-maddocks-royal-festival-hall#:~:text=But%20if%20the%20music%20is,I'm%20at%20it.%E2%80%9D.

47. madison moore, *Fabulous: The Rise of the Beautiful Eccentric* (New Haven, CT: Yale University Press, 2018), 4.

Bibliography

Abbate, Carolyn. "Music—Drastic or Gnostic?" *Critical Inquiry* 30, no. 3 (Spring 2004): 505–36.
Abel, Sam. *Opera in the Flesh: Sexuality in Operatic Performance.* Boulder, CO: Westview Press, 1996.
Ablow, Rachel. *The Marriage of Minds: Reading Sympathy in the Victorian Marriage Plot.* Stanford, CA: Stanford University Press, 2007.
The Adventures of a School-Boy. In *The New Epicurean and The Adventures of a School-Boy: Two Tales from the Victorian Underground*, 97–219. New York: Grove Press, 1984.
Advertisement for the dactylergon. *English Mechanic and World of Science*, no. 1295 (January 17, 1890): 432.
Advertisement for the technicon. *Monthly Musical Record* (1 May 1887): 119.
Ahmed, Sara. *The Cultural Politics of Emotion.* London: Routledge, 2004.
———. "Open Forum: Imaginary Prohibitions; Some Preliminary Remarks on the Founding Gestures of the 'New Materialism.'" *European Journal of Women's Studies* 15, no. 1 (2008): 23–39.
———. *Queer Phenomenology: Orientations, Objects, Others.* Durham, NC: Duke University Press, 2006.
———. *What's the Use?* Durham, NC: Duke University Press, 2019.
Allen, Grant. *Physiological Aesthetics.* London: Henry S. King, 1877.
Amin, Kadji. *Disturbing Attachments: Genet, Modern Pederasty, and Queer History.* Durham, NC: Duke University Press, 2017.
Anolik, Ruth Bienstock. "Sexual Horror: Fears of the Sexual Other." In *Horrifying Sex: Essays on Sexual Deviance in Gothic Literature*, edited by Ruth Bienstock Anolik, 1–26. Jefferson, NC: McFarland, 2007.
Anthes, Emily. "Musical Chairs? Swapping Seats Could Reduce Orchestra Aerosols." *New York Times*, June 23, 2021. https://www.nytimes.com/2021/06/23/health/coronavirus-orchestra.html.
Argyle, Gisela. "Mrs. Humphry Ward's Fictional Experiments in the Woman Question." *Studies in English Literature, 1500–1900* 43, no. 4 (Autumn 2003): 939–57.

Armstrong, Isobel. "Bodily Things and Thingly Bodies: Circumventing the Subject-Object Binary." In *Bodies and Things in Nineteenth-Century Literature and Culture*, edited by Katharina Boehm, 17–44. London: Palgrave Macmillan, 2012.

Armstrong, Tim. "Hardy, History, and Recorded Music." In *Thomas Hardy and Contemporary Literary Studies*, edited by Tim Dolin and Peter Widdowson, 153–66. London: Palgrave Macmillan, 2004.

Arnim, Anna Leffler. *A Complete Course of Wrist and Finger Gymnastic: For Students of the Piano, Organ, Violin, and Other Instruments*. 3rd ed. London: Hutchings and Crowsley, 1894.

Ashton, Rosemary. *George Eliot, A Life*. London: Penguin, 1997.

Asquith, Mark. *Thomas Hardy, Metaphysics, and Music*. Hampshire, UK: Palgrave Macmillan, 2005.

Auerbach, Emily. *Maestros, Dilettantes, and Philistines: The Musician in the Victorian Novel*. Bern: Peter Lang, 1989.

Auslander, Philip. "Musical Personae." *TDR* 50, no. 1 (2006): 100–19.

Auyoung, Elaine. "Phantoms and Fictional Persons: Hardy's Phenomenology of Loss." *Victorian Studies* 59, no. 3 (Spring 2017): 399–408.

Bagocius, Benjamin. "Masturbation and Physiological Romance in Teleny." *Criticism* 59, no. 3 (2017): 441–67.

"Baiser." *Le Petit Robert de la Langue Française*. Accessed June 13, 2019, https://dictionnaire.lerobert.com/definition/baiser.

Baitz, Dana. "Toward a Trans* Method in Musicology." In *The Oxford Handbook of Music and Queerness*, edited by Fred Everett Maus and Sheila Whiteley, 367–82. Oxford: Oxford University Press, 2022.

Baker, Frederick Charles. *How We Hear: A Treatise on the Phenomena of Sound*. London: Vincent Music, 1901.

Barnard, Charles. *Camilla, The Tale of a Violin, Being the Artist Life of Camilla Urso*. Boston: Loring, 1874.

Bartlett, Neil. *Who Was That Man? A Present for Mr. Oscar Wilde*. London: Serpent's Tail, 1988.

Barton, Ruth. "'An Influential Set of Chaps': The X-Club and Royal Society Politics, 1865–85." *British Journal for the History of Science* 23, no. 1 (March 1990): 53–81.

Bashant, Wendy. "Singing in Greek Drag: Gluck, Berlioz, George Eliot." In *En Travesti: Women, Gender Subversion, Opera*, edited by Corinne E. Blackmer and Patricia Juliana Smith, 216–41. New York: Columbia University Press, 1995.

Bashford, Christina. "Sensuality and the Senses: The Appeal of the Violin in Victorian Culture." Conference paper presented at the Midwest Victorian Studies Association, Iowa City, Iowa, May 1, 2015.

Bates, Eliot. "The Social Life of Musical Instruments." *Ethnomusicology* 56, no. 3 (Fall 2012): 363–95.

Baughan, Edward. "Marchesi and Singing." *Monthly Musical Record* 28, no. 325 (1898): 7–8.

Baumgardner, Astrid. "Engaging Today's Audience: 3 Things That Can Make a Difference to Classical Music." *Astrid Baumgardner* (blog), November 29, 2012. https://www.astridbaumgardner.com/blog-and-resources/blog/engaging-todayu2019s-audience-3-things-that-can-make-a-difference-to-classical-music/.

Beardsley, Aubrey. *The Story of Venus and Tannhäuser: A Romantic Novel*. For Private Circulation, 1907.

Beatty, Jackson, and Brennis Lucero-Wagoner. "The Pupillary System." In *Handbook of Psychophysiology*, edited by John T. Cacioppo, Louis G. Tassinary, and Gary G. Berntson, 142–62. Cambridge: Cambridge University Press, 2000.

Beer, Gillian. *George Eliot*. Brighton, UK: Harvester Press, 1986.

———. *Open Fields: Science in Cultural Encounter*. Oxford: Oxford University Press, 1996.

Behar, Katherine. "An Introduction to OOF." In *Object-Oriented Feminism*, edited by Katherine Behar, 1–36. Minneapolis: University of Minnesota Press, 2016.

Bennett, James, II. "What's Up with All Those Ricola Cough Drops at Carnegie Hall?" *WQXR: How to Classical*, June 6, 2018. https://www.wqxr.org/story/ricola-cough-drops-carnegie-hall/.

Bennett, Jane. *Vibrant Matter: A Political Ecology of Things*. Durham, NC: Duke University Press, 2010.

Bernstein, Susan. "On Music Framed: The Eolian Harp in Romantic Writing." In *The Figure of Music in Nineteenth-Century British Poetry*, edited by Phyllis Weliver, 70–84. London: Routledge, 1995.

Bessette, Jean. "Queer Rhetoric in Situ." *Rhetoric Review* 35, no. 2 (2016): 148–64.

Best, Stephen, and Sharon Marcus. "Surface Reading: An Introduction." *Representations* 108, no. 1 (Fall 2009): 1–21.

Billroth, Theodor. *Wer ist musikalisch?* Berlin: Gebrüder Paetel, 1895.

Boehm, Katharina. "Introduction: Bodies and Things." In *Bodies and Things in Nineteenth-Century Literature and Culture*, edited by Katharina Boehm, 1–16. London: Palgrave Macmillan, 2012.

Booth, Alison. "Feminism." *Victorian Literature and Culture* 46, no. 3/4 (Fall/Winter 2018): 691–97.

Bosanquet, R. H. M. *An Elementary Treatise on Musical Intervals and Temperament*. London: Macmillan, 1876.

Bowers, Jane, and Judith Tick. Introduction to *Women Making Music: The Western Art Tradition, 1150–1950*, edited by Jane Bowers and Judith Tick, 3–14. Urbana: University of Illinois Press, 1986.

Bowman, Wayne D. *Philosophical Perspectives on Music*. Oxford: Oxford University Press, 1998.

Bradway, Tyler, and Elizabeth Freeman. "Introduction: Kincoherence/Kin-aesthetics/Kinematics." In *Queer Kinship: Race, Sex, Belonging, Form*, edited by Tyler Bradway and Elizabeth Freeman, 1–24. Durham, NC: Duke University Press, 2022.

Braidotti, Rosi. *Posthuman Feminism*. Cambridge: Polity, 2022.

Brain, Robert Michael. *The Pulse of Modernism: Physiological Aesthetics in Fin-de-Siècle Europe.* Seattle: University of Washington Press, 2015.

Brett, Phillip. "Musicality, Essentialism, and the Closet." In *Queering the Pitch*, edited by Philip Brett and Elizabeth Wood, 9–26. New York: Routledge, 1994.

Bristow, Joseph. "'A Few Drops of Thick, White, Viscid Sperm': *Teleny* and the Defense of the Phallus." In *Porn Archives*, edited by Tim Dean, Steven Ruszczycky, and David Squires, 144–60. Durham, NC: Duke University Press, 2014.

Broadhouse, John. Preface to *The Student's Helmholtz: Musical Acoustics; or, The Phenomena of Sound as Connected with Music*, by John Broadhouse, vii–ix. London: William Reeves, 1881.

Broughton, Rhoda. "The Truth, the Whole Truth, and Nothing but the Truth." In *The Oxford Book of Victorian Ghost Stories*, edited by Michael Cox and R. A. Gilbert, 74–82. Oxford: Oxford University Press, 2003.

Brown, Clive. "Louis Spohr." *Grove Music Online*, Oxford University Press, 2001. https://www.oxfordmusiconline.com/grovemusic/view/10.1093/gmo/9781561592630.001.0001/omo-9781561592630-e-0000026446.

Bucknell, Brad. *Literary Modernism and Musical Aesthetics: Pater, Pound, Joyce, and Stein.* Cambridge: Cambridge University Press, 2001.

Burchfield, J. D. "John Tyndall at the Royal Institution." In *"The Common Purposes of Life": Science and Society at the Royal Institution of Great Britain*, edited by Frank A. J. L. James, 147–68. London: Routledge, 2017.

Burdett, Carolyn. "Sexual Selection, Automata and Ethics in George Eliot's *The Mill on the Floss* and Olive Schreiner's *Undine* and *From Man to Man*." *Journal of Victorian Culture* 14, no. 1 (January 2009): 26–52.

———. "'The Subjective Inside Us Can Turn Into the Objective Outside': Vernon Lee's Psychological Aesthetics." *19: Interdisciplinary Studies in the Long Nineteenth Century* 12 (2011). https://19.bbk.ac.uk/article/id/1549/.

Burgan, Mary. "Heroines at the Piano: Women and Music in Nineteenth-Century Fiction." *Victorian Studies* 30, no. 1 (Autumn 1986): 51–76.

Burgh, Allaston. *Anecdotes of Music, Historical and Biographical; in a Series of Letters from a Gentleman to His Daughter.* London: Longman, Hurst, Rees, Orme, and Brown, 1814.

Burke, Edmund. *A Philosophical Enquiry into the Origin of Our Ideas of the Sublime and Beautiful*, edited by James T. Boulton. South Bend, IN: University of Notre Dame Press, 1986.

Burstein, Miriam Elizabeth. Introduction to *Robert Elsmere*, edited by Miriam Elizabeth Burstein, 5–11. Brighton, UK: Victorian Secrets, 2013.

Burton, Nigel. "Florence Ashton Marshall." *Grove Music Online*, Oxford University Press, 2001. https://doi.org/10.1093/gmo/9781561592630.article.2020251.

Buse, Peter, and Andrew Stott. "Introduction: A Future for Haunting." In *Ghosts: Deconstruction, Psychoanalysis, History*, edited by Peter Buse and Andrew Stott, 1–20. London: Palgrave Macmillan, 1999.

Butler, Judith. *Bodies That Matter: On the Discursive Limits of "Sex."* New York: Routledge, 2011.
———. *Gender Trouble: Feminism and the Subversion of Identity.* New York: Routledge, 2006.
———. "Is Kinship Always Already Heterosexual?" *differences* 13, no. 1 (2002): 14–44.
———. *Undoing Gender.* New York: Routledge, 2004.
Byerly, Alison. *Realism, Representation, and the Arts in Nineteenth-Century Literature.* Cambridge: Cambridge University Press, 1997.
Caballero, Carlo. "'A Wicked Voice': On Vernon Lee, Wagner, and the Effects of Music." *Victorian Studies* 35, no. 4 (Summer 1992): 385–408.
Cahan, David. *Helmholtz: A Life in Science.* Chicago: University of Chicago Press, 2018.
———. "Helmholtz and the British Scientific Elite: From Force Conversation to Energy Conservation." *Royal Society Journal of the History of Science* 66 (2012): 55–68.
Caleb, Amanda Mordavsky, ed. Introduction to *Teleny; or, the Reverse of the Medal.* Richmond, VA: Valancourt Books, 2010. Kindle.
Campbell, Daisy Rhodes. *The Violin Lady.* Boston: Page Company, 1916.
Capes, Bernard. "An Eddy on the Floor." In *The Oxford Book of Victorian Ghost Stories*, edited by Michael Cox and R. A. Gilbert, 403–30. Oxford: Oxford University Press, 2003.
Carey, E. "Body and Music." *National Review* 5, no. 27 (May 1885): 382–90.
Castle, Terry. *The Apparitional Lesbian: Female Homosexuality and Modern Culture.* New York: Columbia University Press, 1993.
Charlwood, Catherine. "'Habitually Embodied' Memories: The Materiality and Physicality of Music in Hardy's Poetry." *Nineteenth-Century Music Review* 17 (2020): 245–69.
Chatterjee, Ronjaunee, Alicia Mireles Christoff, and Amy R. Wong. "Introduction: Undisciplining Victorian Studies." *Victorian Studies* 62, no. 3 (Spring 2020): 369–91.
Chauncey, George. *Gay New York: Gender, Urban Culture, and the Making of the Gay Male World, 1890–1940.* New York: Basic Books, 1994.
Cheng, William. *Just Vibrations: The Purpose of Sounding Good.* Ann Arbor: University of Michigan Press, 2016.
———. *Loving Music Till It Hurts.* Oxford: Oxford University Press, 2019.
Chieffo, Gene. "Blind Symphonic Auditions: Mitigating the Body's Effect on Orchestral Jobs." *Bodylore* (blog), March 4, 2020. https://sites.wp.odu.edu/bodylore/2020/03/04/blind-symphonic-auditions-mitigating-the-bodys-effect-on-orchestral-jobs/.
Chivers, Meredith, and Lori Brotto. "Controversies of Women's Sexual Arousal and Desire." *European Psychologist* 22, no. 1 (2017): 5–26.

Chomet, Hector. *The Influence of Music on Health and Life.* Translated by Laura A. Flint. New York: G. P. Putnam's Sons, 1875.

Chow, Jeremy. "Mellifluent Sexuality: Female Intimacy in Ann Radcliffe's *The Romance of the Forest.*" *Eighteenth-Century Fiction* 30, no. 2 (Winter 2017–18): 195–221.

Chua, Daniel. *Absolute Music and the Construction of Meaning.* Cambridge: Cambridge University Press, 1999.

Clapp-Itnyre, Alisa. *Angelic Airs, Subversive Songs: Music as Social Discourse in the Victorian Novel.* Athens: Ohio University Press, 2002.

Clark, Nick. "Cacophony of Sound? People Cough on Purpose During Classical Concerts, Research Finds." *The Independent,* January 29, 2013. https://www.independent.co.uk/arts-entertainment/classical/news/cacophony-sound-people-cough-purpose-during-classical-concerts-research-finds-8471735.html.

Clayton, Jay. "Hacking the Nineteenth Century." In *Victorian Afterlife: Postmodern Culture Rewrites the Nineteenth Century,* edited by John Kucich and Dianne F. Sadoff, 186–210. Minneapolis: University of Minnesota Press, 2000.

Cohen, Cathy. "Punks, Bulldaggers, and Welfare Queens: The Radical Potential of Queer Politics?" *GLQ* 3, no. 4 (May 1997): 437–65.

Cohen, Ed. "Writing Gone Wilde: Homoerotic Desire in the Closet of Representation." *PMLA* 102, no. 5 (October 1987): 801–13.

Cohen, William. *Embodied: Victorian Literature and the Senses.* Minneapolis: University of Minnesota Press, 2008.

———. *Sex Scandal: The Private Parts of Victorian Fiction.* Durham, NC: Duke University Press, 1996.

Cohn, Elisha. *Still Life: Suspended Development in the Victorian Novel.* Oxford: Oxford University Press, 2016.

Colby, Vineta. *Vernon Lee: A Literary Biography.* Charlottesville: University of Virginia Press, 2003.

Colligan, Collette. *A Publisher's Paradise: Expatriate Literary Culture in Paris, 1890–1960.* Amherst: University of Massachusetts Press, 2013.

———. *The Traffic in Obscenity from Byron to Beardsley: Sexuality and Exoticism in Nineteenth-Century Print Culture.* London: Palgrave Macmillan, 2006.

"Concert Etiquette." *Dwight's Journal of Music* 24, no. 4 (May 14, 1864): 234.

Cook, Matt. *London and the Culture of Homosexuality, 1885–1914.* Cambridge: Cambridge University Press, 2003.

Cook, Nicholas. *Beyond the Score: Music as Performance.* Oxford: Oxford University Press, 2013.

Coombs, David Sweeney. *Reading with the Senses in Victorian Literature and Science.* Charlottesville: University of Virginia Press, 2019.

"Correction of Faulty Touch." *Musical Herald,* no. 493 (April 1889): 82.

Cox, Kimberly. *Touch, Sexuality, and Hands in British Literature, 1740–1901.* London: Routledge, 2022.

Cox, Michael, and R. A. Gilbert. Introduction to *The Oxford Book of Victorian Ghost Stories*, edited by Michael Cox and R. A. Gilbert, ix–xx. Oxford: Oxford University Press, 2003.

Crary, Jonathan. *Suspensions of Perception: Attention, Spectacle, and Modern Culture*. Cambridge, MA: MIT Press, 1999.

Craster, Edmuind. "Personal Note on John Meade Falkner." In *The Nebuly Coat and The Lost Stradivarius*, xi–xii. Oxford: Oxford University Press, 1954.

Curtis, John Harrison. *On the Cephaloscope and Its Uses in the Discrimination of the Normal and Abnormal Sounds in the Organ of Hearing*. London: John Churchill, 1842.

———. *A Treatise on the Physiology and Diseases of the Ear*. London: John Anderson, 1819.

Cusick, Suzanne. "Feminist Theory, Music Theory, and the Mind/Body Problem." *Perspectives of New Music* 32, no. 1 (Winter 1994): 8–27.

———. "Musicology, Torture, Repair." *Radical Musicology* 3 (2008). http://www.radical-musicology.org.uk/2008/Cusick.htm.

———. "On a Lesbian Relationship with Music: A Serious Effort Not to Think Straight." In *Queering the Pitch*, edited by Philip Brett and Elizabeth Wood, 67–83. New York: Routledge, 1994.

———. "On Musical Performances of Sex and Gender." In *Audible Traces: Gender, Identity, and Music*, edited by Elaine Barkin and Lydia Hamessley, 25–49. Zurich: Carciofoli Verlagshaus, 1999.

———. "Response: 'This Song Is for You.'" *Journal of the American Musicological Society* 66, no. 3 (Fall 2013): 861–72.

da Sousa Correa, Delia. *George Eliot, Music and Victorian Culture*. London: Palgrave Macmillan, 2003.

———. "Music." In *George Eliot in Context*, edited by Margaret Harris, 206–13. Cambridge: Cambridge University Press, 2013.

Dalhaus, Carl. *The Idea of Absolute Music*. Translated by Roger Lustig. Chicago: University of Chicago Press, 1989.

Dalley, Lana, and Kellie Holzer. "Victorian Literature in the Age of #MeToo: An Introduction." *Nineteenth-Century Gender Studies* 16, no. 2 (Summer 2020). http://ncgsjournal.com/issue162/introduction.html.

Daly, Nicholas. "Somewhere There's Music: John Meade Falkner's *The Lost Stradivarius*." In *Romancing Decay: Ideas of Decadence in European Culture*, edited by Michael St. John, ch. 7. New York: Routledge, 2016. Google Books.

Dames, Nicholas. *The Physiology of the Novel: Reading, Neural Science, and the Form of Victorian Fiction*. Oxford: Oxford University Press, 2007.

Daston, Lorraine, and Peter Galison. *Objectivity*. Brooklyn: Zone Books, 2007.

Davies, James Q. *Romantic Anatomies of Performance*. Berkeley: University of California Press, 2014.

Davis, Noela. "New Materialism and Feminism's Anti-biologism: A Response to Sara Ahmed." *European Journal of Women's Studies* 16, no. 1 (2009): 67–80.

Davis, William. "Music Therapy in Victorian England." *British Journal of Music Therapy* 2, no. 1 (1988): 10–16.

Davison, Alan. "High-Art Music and Low-Brow Types: Physiognomy and Nineteenth-Century Music Iconography." *Context* 17 (Winter 1999): 5–19.

Dean, Tim. *Beyond Sexuality*. Chicago: University of Chicago Press, 2000.

———. "The Biopolitics of Pleasure." *South Atlantic Quarterly* 111, no. 3 (2012): 477–96.

Delgado, Maria M. "Carles Santos, 'Music in the Theatre.'" In *Taking It to the Bridge: Music as Performance*, edited by Nicholas Cook and Richard Pettengill, 237–61. Ann Arbor: University of Michigan Press, 2013.

Dellamora, Richard. "Male Relations in Thomas Hardy's *Jude the Obscure*." *Papers on Language and Literature* 27, no. 4 (Fall 1991): 453–72.

Der Bedrosian, Jeanette. "Rock On—Just Not Too Hard." *Johns Hopkins Magazine*, Summer 2019. https://hub.jhu.edu/magazine/2019/summer/guitar-injury-smartguitar-serap-bastepe-gray/.

Deutsch, David. *British Literature and Classical Music: Cultural Contexts, 1870–1945*. London: Bloomsbury, 2015.

Dickens, Charles. *Dombey and Son*. Edited by Andrew Sanders. London: Penguin Classics, 2002.

Dinshaw, Carolyn. *Getting Medieval: Sexualities and Communities, Pre- and Postmodern*. Durham, NC: Duke University Press, 1999.

Donelan, James. *Poetry and the Romantic Musical Aesthetic*. Cambridge: Cambridge University Press, 2009.

"Dr. Swinderton Heap." *Musical Herald*, no. 547 (October 2, 1893): 293.

"Dr. W.H. Stone on the Causes of Rise in Orchestral Pitch and the Methods of Obviating It." *Monthly Musical Record* 11 (May 1, 1881): 90.

Draucker, Shannon. "Food for the Soul, Art of the Flesh: Classical Music, COVID-19, and the Body." *BLARB* (blog), *Los Angeles Review of Books*, September 2, 2021. https://blog.lareviewofbooks.org/essays/food-soul-art-flesh-classical-music-covid-19-body/.

———. "Hearing, Sensing, Feeling Sound: On Music and Physiology in Victorian England, 1857–1894." *BRANCH: Britain, Representation and Nineteenth-Century History*. Edited by Dino Franco Felluga. Extension of *Romanticism and Victorianism on the Net*. https://branchcollective.org/?ps_articles=shannon-draucker-hearing-sensing-feeling-sound-on-music-and-physiology-in-victorian-england-1857-1894.

———. "Ladies' Orchestras and Music-as-Performance in *Fin-de-Siècle* Britain." *Nineteenth-Century Contexts* 45, no. 1 (2023): 7–22.

———. "Music Physiology, Erotic Encounters, and Queer Reading Practices in *Teleny*." *Victorian Literature and Culture* 50, no. 1 (Spring 2022): 141–72.

———. "Performing Power: Female Musicianship and Embodied Artistry in Bertha Thomas's *The Violin-Player*." *Nineteenth-Century Gender Studies* 14, no. 1 (Spring 2018). https://www.ncgsjournal.com/issue141/draucker.html.
Dubourg, George. *The Violin: Some Account of That Leading Instrument and Its Most Eminent Professors*. London: Robert Cocks, 1836.
Eagleton, Terry. *The Ideology of the Aesthetic*. Oxford: Basil Blackwell, 1990.
Earle, Bo. "Policing and Performing Liberal Individuality in Anthony Trollope's *The Warden*." *Nineteenth-Century Literature* 61, no. 1 (June 2006): 1–31.
Edelman, Lee. *No Future: Queer Theory and the Death Drive*. Durham, NC: Duke University Press, 2004.
Ehnenn, Jill. "From 'We Other Victorians' to 'Pussy Grabs Back': Thinking Gender, Thinking Sex, and Feminist Methodological Futures in Victorian Studies." *Victorian Literature and Culture* 41, no. 1 (2019): 35–62.
Eidsheim, Nina Sun. *The Race of Sound: Listening, Timbre, & Vocality in African American Music*. Durham, NC: Duke University Press, 2019.
———. *Sensing Sound: Singing and Listening as Vibrational Practice*. Durham, NC: Duke University Press, 2015.
Eling, Paul, Stanley Finger, and Harry Whitaker, "On the Origins of Organology: Franz Joseph Gall and a Girl Named Bianchi." *Cortext* 86 (2017): 123–31.
Eliot, George. *Daniel Deronda*. Edited by Terence Cave. London: Penguin, 1995.
———. *The George Eliot Letters*. Edited by Gordon S. Haight, vol. 6. New Haven, CT: Yale University Press, 1954.
———. *George Eliot's Daniel Deronda Notebooks*. Edited by Jane Irwin. Cambridge: Cambridge University Press, 1996.
———. *George Eliot's Life as Related in Her Letters and Journals*. Edited by J. W. Cross, vol. 1. Edinburgh: William Blackwood and Sons, 1885.
———. "Liszt, Wagner, and Weimar." *Fraser's Magazine* 52 (July 1855): 48–62.
———. "Mr. Gilfil's Love Story." In *Scenes of Clerical Life*, edited by Jennifer Gribble, 77–194. London: Penguin Books, 1998.
———. *Middlemarch*. Edited by Rosemary Ashton. London: Penguin Books, 1994.
———. *The Mill on the Floss*. Edited by A. S. Byatt. London: Penguin Books, 2003.
———. "O May I Join the Choir Invisible." In *The Legend of Jubal and Other Poems*, 231–33. Boston: James Osgood, 1874.
———. *Romola*. Edited by Dorothea Barrett. London: Penguin Books, 1996.
———. "Stradivarius." In *The Legend of Jubal and Other Poems*, 205–16. Boston: James R. Osgood, 1874.
Ellis, Havelock. *Studies in the Psychology of Sex*. Vol. 2, 3rd ed. Philadelphia: F. A. Davis, 1915.
———. *Studies in the Psychology of Sex*. Vol. 4. Philadelphia: F. A. Davis, 1905.
Engelstein, Stefani. *Sibling Action: The Genealogical Structure of Modernity*. New York: Columbia University Press, 2007.

Epstein, Josh. *Sublime Noise: Musical Culture and the Modernist Writer.* Baltimore: Johns Hopkins University Press, 2014.
Esmail, Jennifer. *Reading Victorian Deafness: Signs and Sounds in Victorian Literature and Culture.* Athens: Ohio University Press, 2013.
"An Evening with the Royal Amateur Orchestral Society." *Musical Standard* 34, no. 19 (February 1888): 98.
Ewell, Philip. *On Music Theory, and Making Music More Welcoming for Everyone.* Ann Arbor: University of Michigan Press, 2023.
F. G. E. "Lady Violinists (Concluded)." *Musical Times* 47, no. 765 (November 1, 1906): 739.
"The Fair Sex-Tett." *Punch* 68 (April 3, 1875): 150.
Falkner, John Meade. *The Lost Stradivarius.* New York: D. Appleton, 1896.
"Female Orchestral Players." *Musical News* 4, no. 112 (April 22, 1893): 372.
Fennell, James. "On the Effects of Music on Man and Animals." *Mirror of Literature, Amusement, and Instruction* 38, no. 1081 (October 16, 1841): 245–48.
Fidler, Florence G. "Women as Orchestral Players." *Musical News* 38, no. 474 (March 31, 1900): 310.
Forster, E. M. "Dr. Woolacott." In *The Life to Come and Other Stories.* New York: Norton, 1972.
Foucault, Michel. *The History of Sexuality.* Vol. I. Translated by Robert Hurley. New York: Vintage Books, 1990.
Francis, M. E. *The Duenna of a Genius.* Leipzig: Bernhard Tauchnitz, 1899.
Franklin, Sarah, and Susan McKinnon. Introduction to *Relative Values: Reconfiguring Kinship Studies,* edited by Sarah Franklin and Susan McKinnon, 1–28. Durham, NC: Duke University Press, 2002.
Freccero, Carla. *Queer/Early/Modern.* Durham, NC: Duke University Press, 2006.
Freedgood, Elaine. *The Ideas in Things: Fugitive Meaning in the Victorian Novel.* Chicago: University of Chicago Press, 2006.
Freeman, Elizabeth. *Time Binds: Queer Temporalities, Queer Histories.* Durham, NC: Duke University Press, 2010.
Freitas, Roger. "The Eroticism of Emasculation: Confronting the Baroque Body of the Castrato." *Journal of Musicology* 20, no. 2 (Spring 2003): 196–249.
Fretwell, Erica. *Sensory Experiments: Psychophysics, Race, and the Aesthetics of Feeling.* Durham, NC: Duke University Press, 2020.
Fricke, Christel. "Kant." In *Music in German Philosophy: An Introduction,* edited by Stefan Lorenz Sorgner and Oliver Fürbeth. Translated by Susan H. Gillespie, 27–46. Chicago: University of Chicago Press, 2010.
Friedman, David. *Wilde in America: Oscar Wilde and the Invention of Modern Celebrity.* New York: W. W. Norton, 2014.
Friedman, Dustin. *Before Queer Theory: Victorian Aestheticism and the Self.* Baltimore: Johns Hopkins University Press, 2019.
Frisch, Walter. *Music in the Nineteenth Century.* New York: Norton, 2012.

Fuller, Sophie. "Dead White Men in Wigs." In *Girls! Girls! Girls! Essays on Women and Music*, edited by Sarah Cooper, 22–36. New York: New York University Press, 1996.
Furneaux, Holly. *Queer Dickens: Erotics, Families, Masculinities*. Oxford: Oxford University Press, 2013.
Fuss, Diana. *Dying Modern: A Meditation on Elegy*. Durham, NC: Duke University Press, 2013.
Gagné, Ann. *Embodying the Tactile in Victorian Literature: Touching Bodies/Bodies Touching*. Lanham, MD: Rowman and Littlefield, 2021.
Gall, Franz Joseph. *On the Functions of the Brain and Each of Its Parts*. Vol. 1. Translated by Winslow Lewis. Boston: Marsh, Capen & Lyon, 1835.
———. *On the Functions of the Brain and Each of Its Parts*. Vol. 5. Translated by Winslow Lewis. Boston: Marsh, Capen & Lyon, 1835.
Gallagher, Catherine, and Stephen Greenblatt. *Practicing New Historicism*. Chicago: University of Chicago Press, 2001.
Gatrell, Simon. *Thomas Hardy's Vision of Wessex*. New York: Palgrave Macmillan, 2003.
Gillett, Paula. *Musical Women in England, 1870–1914: "Encroaching on All Man's Privileges."* New York: St. Martin's, 2000.
Gillin, Edward J. *Sound Authorities: Scientific and Musical Knowledge in Nineteenth-Century Britain*. Chicago: University of Chicago Press, 2021.
Gjerris, Helene, and Bent Nørgaard. "Improving the Performance of Classical Musicians through Interdisciplinary Exercises." *Arts and Humanities as Higher Education*, August 2016. http://www.artsandhumanities.org/improving-the-performance-of-classical-musicians-through-interdisciplinary-exercises/.
Glover, Robin Bickerstaff. "Symphony Etiquette." Maryland Symphony Orchestra. https://www.marylandsymphony.org/plan-your-visit/what-expect.
Golding, Rosemary. *Music and Academia in Victorian Britain*. London: Routledge, 2016.
———. *Music and Moral Management in the Nineteenth-Century English Lunatic Asylum*. London: Palgrave Macmillan, 2021.
Gooley, Dana. "The Battle Against Instrumental Virtuosity in the Early Nineteenth Century." In *Franz Liszt and His World*, edited by Christopher H. Gibbs and Dana Gooley, 75–112. Princeton, NJ: Princeton University Press, 2006.
Gossin, Pamela. *Thomas Hardy's Novel Universe: Astronomy, Cosmology, and Gender in the Post-Darwinian World*. Aldershot: Ashgate, 2007.
Gould, Stephen Jay. *The Mismeasure of Man*. New York: W. W. Norton, 1996.
Grand, Sarah. *The Heavenly Twins*. Edited by Carol Senf. Ann Arbor: University of Michigan Press, 1994.
Grant, Roger Matthew. "Peculiar Attunements: Comic Opera and Enlightenment Mimesis." *Critical Inquiry* 43, no. 2 (2017): 550–69.
———. *Peculiar Attunements: How Affect Theory Turned Musical*. New York: Fordham University Press, 2020.

Gray, Beryl. *George Eliot and Music*. New York: Palgrave Macmillan, 1989.
Green, Jeffrey. *Samuel Coleridge-Taylor, a Musical Life*. London: Routledge, 2011.
Green, Lucy. *Music, Gender, Education*. Cambridge: Cambridge University Press, 1997.
Greiner, Rae. *Sympathetic Realism in Nineteenth-Century British Fiction*. Baltimore, MD: Johns Hopkins University Press, 2012.
Grosz, Elizabeth. *Time Travels: Feminism, Nature, Power*. Durham, NC: Duke University Press, 2005.
———. *Volatile Bodies: Toward a Corporeal Feminism*. Bloomington: Indiana University Press, 1994.
Grundy, Joan. *Hardy and the Sister Arts*. London: Macmillan, 1979.
Gurman, Elissa. "Sex, Consent, and the Unconscious Female Body: Reading *Tess of the d'Urbervilles* alongside the Trial of Brock Allen Turner." *Law and Literature* 32, no. 1 (2019): 155–70.
Gurney, Edmund. "On Some Disputed Points in Music." *Fortnightly Review* 20, no. 115 (1876): 106–30.
———. *The Power of Sound*. London: Smith, Elder, 1880.
Haefele-Thomas, Ardel. "Introduction: Trans Victorians." *Victorian Review* 44, no. 1 (Spring 2018): 31–36.
Hager, Lisa. "A Case for a Trans Studies Turn in Victorian Studies: 'Female Husbands' of the Nineteenth Century." *Victorian Review* 44, no. 1 (Spring 2018): 37–54.
Halberstam, Jack. *In a Queer Time and Place: Transgender Bodies, Subcultural Bodies*. New York: New York University Press, 2005.
———. *Skin Shows*. Durham, NC: Duke University Press, 1995.
Halperin, David. *Saint Foucault: Towards a Gay Hagiography*. Oxford: Oxford University Press, 1995.
Hamner, L., et al. "High SARS-CoV-2 Attack Rate Following Exposure at a Choir Practice—Skagit County, Washington, March 2020." *Morbidity and Mortality Weekly Report* 69, no. 19 (May 15, 2020): 606–10.
Hardy, Thomas. *The Collected Letters of Thomas Hardy*. Edited by Richard Little Purdy and Michael Millgate. Oxford: Clarendon Press, 1978.
———. *Desperate Remedies*. Edited by Patricia Ingham. Oxford: Oxford University Press, 2003.
———. *Far from the Madding Crowd*. Edited by Rosemary Morgan and Shannon Russell. London: Penguin, 2003.
———. "The Fiddler of the Reels." In *Life's Little Ironies*, edited by Simon Gatrell, 137–55. Oxford: Oxford University Press, 2008.
———. "Haunting Fingers." In *Thomas Hardy: The Complete Poems*, edited by James Gibson, 590–92. London: Palgrave Macmillan, 2001.
———. *Literary Notebooks of Thomas Hardy*. 2 vols. Edited by Lennart A. Björk. London: Macmillan, 1985.
———. *The Return of the Native*. Edited by Simon Avery. Toronto: Broadview, 2013.

———. *Tess of the D'Urbervilles*. Edited by Simon Gatrell. Oxford: Oxford World's Classics, 2008.

———. *Under the Greenwood Tree*. Edited by Claire Seymour. London: Wordsworth, 1994.

Haritaworn, Jinthana. "Decolonizing the Non/Human." *GLQ* 21, nos. 2–3 (June 2015): 210–13.

Harris, T. F. *Hand Book of Acoustics for the Use of Musical Students*. 3rd ed. London: J. Curwen and Sons, 1887.

Hartley, David. *Observations on Man, His Frame, His Duty, and His Expectations*. 6th ed. London: Thomas Tegg and Son, 1834.

Haweis, H. R. *My Musical Life*. 8th ed. New York: Longmans, Green, 1912.

Hawgood, Alex. "Hahn-Bin Straddles Classical Music and Fashion." *New York Times*, February 23, 2011. https://www.nytimes.com/2011/02/24/fashion/24HAHN BIN.html.

Heilmann, Ann. *New Woman Strategies: Sarah Grand, Olive Schreiner, and Mona Caird*. Manchester: Manchester University Press, 2004.

Helmholtz, Hermann von. *An Autobiographical Sketch: An Address Delivered on the Occasion of his Jubilee*. Translated by E. Atkinson. London: Longmans, Green, 1898.

———. *On the Sensations of Tone as a Physiological Basis for the Theory of Music*. Translated by Alexander Ellis. 3rd ed. London: Longmans, Green, 1895.

———. *Popular Lectures on Scientific Subjects*. Translated by E. Atkinson. New York: D. Appleton, 1881.

Henchman, Anna. "Sentience." *Victorian Literature and Culture* 46, nos. 3–4 (2018): 861–65.

Henry, Nancy. *The Life of George Eliot: A Critical Biography*. West Sussex: Wiley Blackwell, 2012.

Herschel, John. "Sound." In *Encyclopedia Metropolitana*. London: William Clowes and Sons, 1845.

Hiebert, Elfrieda. "Listening to the Piano Pedal: Acoustics and Pedagogy in Late Nineteenth-Century Contexts." *Osiris* 28, no. 1 (January 2013): 232–53.

Hoffmann, E. T. A. "Beethoven's Instrumental Music." Translated by Arthur Ware Locke. *Musical Quarterly* 3, no. 1 (January 1917): 127–33.

"A Holiday Afternoon with Professor Tyndall." *Daily News*, no. 8,635 (December 29, 1873): 5.

Honeyman, William Crawford. *The Secrets of Violin Playing*. Edinburgh: E. Köhler & Son, 1800s?.

Huffer, Lynne. *Are the Lips a Grave? A Queer Feminist on the Ethics of Sex*. New York: Columbia University Press, 2013.

Hughes, John. *"Ecstatic Sound": Music and Individuality in the Work of Thomas Hardy*. Aldershot: Ashgate, 2001.

Hughes, Linda K. "'Phantoms of Delight: Amy Levy and Romantic Men." In *Decadent Romanticism: 1780–1914*, edited by Kostas Boyiopoulos and Mark Sandy, 161–75. Aldershot: Ashgate, 2015.

Hui, Alexandra. *The Psychophysical Ear: Musical Experiments, Experimental Sounds, 1840–1910*. Cambridge, MA: MIT Press, 2013.

Hunter, J. Ewing. "Is Soft Music a Calmative in Cases of Fever?" *British Medical Journal* 2, no. 1660 (1892): 923.

Huxley, T. H. "On the Hypothesis That Animals Are Automata, and Its History." *Fortnightly Review*, n.s., 16, no. 95 (November 1, 1874): 555–80.

"The Importance of the Nose in Connexion with the Practice of Singing, and Playing Wind Instruments." *Musical World* 32, no. 41 (October 14, 1854): 685.

"The Influence of Music on the Lower Animals." *All the Year Round* 30, no. 735 (December 30, 1882): 538.

Irwin, Michael. *Reading Hardy's Landscapes*. New York: St. Martin's, 2000.

Jackson, Myles W. *Harmonious Triads: Physicists, Musicians, and Instrument Makers in Nineteenth-Century Germany*. Cambridge, MA: MIT Press, 2006.

Jackson-Houlston, C. M. "Thomas Hardy's Use of Traditional Song." *Nineteenth-Century Literature* 44, no. 3 (December 1989): 301–34.

James, David. "Hearing Hardy: Soundscapes and the Profitable Reader." *Journal of Narrative Theory* 40 no. 2 (Summer 2010): 131–55.

Jarvis, Claire. *Exquisite Masochism: Marriage, Sex, and the Novel Form*. Baltimore: Johns Hopkins University Press, 2016.

"Joachim's Rival." *Musical World* 10, no. 58 (March 6, 1880): 153.

Johnson, Barbara. "Muteness Envy." In *The Barbara Johnson Reader: The Surprise of Otherness*, edited by Melissa Feuerstein et al., 200–216. Durham, NC: Duke University Press, 2014.

Johnson, James. *Listening in Paris: A Cultural History*. Berkeley: University of California Press, 1995.

Johnson, Jenny Olivia. "Musical Abjects: Sounds and Objectionable Sexualities." In *The Oxford Handbook of Music and Queerness*, edited by Fred Everett Maus and Sheila Whiteley, 405–20. Oxford: Oxford University Press, 2022.

Johnson, Julian. *Who Needs Classical Music? Cultural Choice and Musical Value*. Oxford: Oxford University Press, 2002.

Jones, Catherine. *Literature and Music in the Atlantic World, 1767–1867*. Edinburgh: Edinburgh University Press, 2014.

Joseph, Abigail. *Exquisite Materials: Episodes in the Queer History of Victorian Style*. Newark: University of Delaware Press, 2019.

Joy, Eileen. "Improbable Manners of Being." *GLQ* 21, nos. 2–3 (June 2015): 221–24.

Joyce, Simon. *LGBT Victorians: Sexuality and Gender in the Nineteenth-Century Archives*. Oxford: Oxford University Press, 2022.

Kahan, Benjamin. *The Book of Minor Perverts*. Chicago: University of Chicago Press, 2019.

Kahl, Benjamin, ed. Introduction to *The Selected Writings of Hermann von Helmholtz*, by Hermann von Helmholtz, xii–xlv. Middletown, CT: Wesleyan University Press, 1971.

Keen, Suzanne. *Thomas Hardy's Brains*. Columbus: Ohio State University Press, 2014.

Keller, Helen. *The Story of My Life*. New York: Doubleday, Page, 1905.

Kennaway, James. "Music and the Body in the History of Medicine." In *The Oxford Handbook of Music and the Body*, edited by Youn Kim and Sander Gilman, 333–48. Oxford: Oxford University Press, 2019.

Kettle, Martin. "Prom 71: BBCSO/Vänskä—Review." September 5, 2013. https://www.theguardian.com/music/2013/sep/05/prom-71-bbcso-vanska-review.

Killeen, Jarlath. "Emptying Time in Anthony Trollope's *The Warden*." In *Victorian Time: Technologies, Standardization, Catastrophes*, edited by Trish Ferguson, 38–56. London: Palgrave Macmillan, 2013.

Kim, Youn, and Sander L. Gilman. "Contextualizing Music and the Body: An Introduction." In *The Oxford Handbook of Music and the Body*, edited by Youn Kim and Sander L. Gilman, 1–20. Oxford: Oxford University Press, 2019.

Kivy, Peter. *Music, Language, and Cognition: And Other Essays in the Aesthetics of Music*. Oxford: Oxford University Press, 2007.

Koestenbaum, Wayne. *The Queen's Throat: Opera, Homosexuality, and the Mystery of Desire*. Boston: Da Capo, 2001.

Krohn-Grimberghe, Lukkas. "We Must Breathe—Why It Is Important to Talk About Race and Racism in the Context of Classical Music." *WXQR* (blog), June 2, 2020. https://www.wqxr.org/story/we-must-breathe/.

Kucich, John. "Curious Dualities: *The Heavenly Twins* (1893) and Sarah Grand's Belated Modernist Aesthetics." In *The New Nineteenth Century: Feminist Readings of Underread Victorian Fiction*, edited by Barbara Leah Harman and Susan Meyer, 195–204. New York: Routledge, 1996.

Latour, Bruno. *Reassembling the Social: An Introduction to Actor-Network Theory*. Oxford: Oxford University Press, 2005.

Law, Joe. "The 'Perniciously Homosexual Art': Music and Homoerotic Desire in *The Picture of Dorian Gray* and Other *Fin-de-Siècle* Fiction." In *The Idea of Music in Victorian Fiction*, edited by Sophie Fuller and Nicky Losseff, 173–96. Aldershot: Ashgate, 2004.

Le Guin, Elisabeth. *Boccherini's Body: An Essay in Carnal Musicology*. Berkeley: University of California Press, 2006.

Lee, Vernon. *The Beautiful: An Introduction to Psychological Aesthetics*. Cambridge: Cambridge University Press, 1913.

———. *Laurus Nobilis: Chapters on Art and Life*. London: John Lane, 1909.

———. *Music and Its Lovers: An Empirical Study of Emotional and Imaginative Responses to Music*. New York: E. P. Dutton, 1933.

———. "A Wicked Voice." In *Hauntings and Other Fantastic Tales*, edited by Catherine Maxwell and Patricia Pulham, 154–81. Peterborough, Ontario: Broadview, 2006.

———. "Winthrop's Adventure." In *A Phantom Lover and Other Dark Tales*, edited by Mike Ashley. London: British Library, 2020. Kindle.

Leighton, Angela. "Ghosts, Aestheticism, and 'Vernon Lee.'" *Victorian Literature and Culture* 28, no. 1 (2000): 1–14.

Levenson, Shirley Frank. "The Use of Music in *Daniel Deronda*." *Nineteenth-Century Fiction* 24, no. 3 (1969): 317–34.

Levine, George. *Reading Thomas Hardy*. Cambridge: Cambridge University Press, 2017.

Levine, Lawrence. *Highbrow/Lowbrow: The Emergence of Cultural Hierarchy in America*. Cambridge, MA: Harvard University Press, 1988.

Levitin, Daniel. *This Is Your Brain on Music: The Science of a Human Obsession*. New York: Penguin, 2006.

Levy, Amy. "Sinfonia Eroica." In *A Minor Poet and Other Verses*, 59–61. London: T. Fisher Unwin, 1884.

Lewes, George Henry. *The Physiology of Common Life*. Vol. 2. New York: D. Appleton, 1875.

Lewis, Helen. "Why I'll Keep Saying 'Pregnant Women.'" *The Atlantic*, October 26, 2021. https://www.theatlantic.com/ideas/archive/2021/10/pregnant-women-people-feminism-language/620468/.

Lindsay, Caroline Blanche Elizabeth. "How to Play the Violin." In *The Girl's Own Indoor Book*, edited by Charles Peters, 172–77. Philadelphia: J. B. Lippincott, 1892.

Lorde, Audre. "Uses of the Erotic: The Erotic as Power." In *Sister Outsider: Essays and Speeches*, 53–59. Berkeley, CA: Crossing, 1984.

Love, Heather. "Doing Being Deviant: Deviance Studies, Description, and the Queer Ordinary." *differences* 26, no. 1 (2015): 74–95.

Lovesey, Oliver. "Religion." In *George Eliot in Context*, edited by Margaret Harris, 238–47. Cambridge: Cambridge University Press, 2013.

Lubey, Kathleen. "Sexual Remembrance in *Clarissa*." *Eighteenth-Century Fiction* 29, no. 1 (Winter 2016–17): 151–78.

———. *What Pornography Knows: Sex and Social Protest Since the Eighteenth Century*. Stanford, CA: Stanford University Press, 2022.

Lunn, Henry C. "Musical Doctors." *Musical Times and Singing-Class Circular* 21, no. 452 (October 1, 1880): 495–96.

Lutz, Deborah. *Pleasure Bound: Victorian Sex Rebels and the New Eroticism*. New York: Norton, 2011.

———. "The Secret Rooms of *My Secret Life*." *English Studies in Canada* 31, no. 1 (March 2005): 118–27.

Lynch, Eve. "Spectral Politics: The Victorian Ghost Story and the Domestic Servant." In *The Victorian Supernatural*, edited by Nicola Bown, Carolyn Burdett, and Pamela Thurschwell, 67–86. Cambridge: Cambridge University Press, 2004.

MacCormack, Patricia. "Queer Posthumanism: Cyborgs, Animals, Monsters, Perverts." In *The Ashgate Research Companion to Queer Theory*, edited by Noreen Giffney and Michael O'Rourke, 111–28. Aldershot: Ashgate, 2009.

MacKinnon, Catharine. *Toward a Feminist Theory of the State*. Cambridge, MA: Harvard University Press, 1989.
Maddocks, Fiona. "Yuja Wang: 'If the Music Is Beautiful and Sensual, Why Not Dress to Fit?'" *The Guardian*, April 9, 2017. https://www.theguardian.com/music/2017/apr/09/yuja-wang-piano-interview-fiona-maddocks-royal-festival-hall#:~:text=But%20if%20the%20music%20is,I'm%20at%20it.%E2%80%9D.
Maddocks, Fiona, and Sasha Valeri Millwood. "Should You Go to a Concert with a Cough?" *The Guardian*, December 6, 2014. https://www.theguardian.com/commentisfree/2014/dec/06/should-you-go-to-concert-cough-kyung-wha-chung.
Mahoney, Kristin. *Queer Kinship After Wilde: Transnational Decadence and the Family*. Cambridge: Cambridge University Press, 2022.
Manne, Kate. *Down Girl: The Logic of Misogyny*. Oxford: Oxford University Press, 2018.
Manning, Céline Frigau. "Phrenologizing Opera Singers: The Scientific 'Proofs of Musical Genius.'" *19th Century Music* 39, no. 2 (Fall 2015): 125–41.
Marcus, Sharon. *Between Women: Friendship, Desire, and Marriage in Victorian England*. Princeton, NJ: Princeton University Press, 2007.
Marcus, Steven. *The Other Victorians: A Study of Sexuality and Pornography in Mid-Nineteenth-Century England*. New York: Basic Books, 1964.
Marsh, Richard. "The Violin." In *The Seen and the Unseen*, 92–115. London: Methuen, 1900.
Martín Alcoff, Linda. *Rape and Resistance*. Cambridge: Polity, 2018.
Mason, Diane. *The Secret Vice: Masturbation in Victorian Fiction and Medical Culture*. Oxford: Oxford University Press, 2006.
Maus, Fred Everett. "'What If Music Is Sex?': Suzanne Cusick and Collaboration." *Radical Musicology* 7 (2019). http://www.radical-musicology.org.uk/2019/Maus.htm.
Maxwell, Catherine. "Sappho, Mary Wakefield, and Vernon Lee's 'A Wicked Voice.'" *Modern Language Review* 102, no. 4 (October 2007): 960–74.
McCartney, R. H. *Barbell or Wand Exercises for Use in Schools with Musical Accompaniment*. London: George Gill and Sons, 1881.
McLaren, Neil, and Rafael Vara Thorbeck. "Little-Known Aspect of Theodor Billroth's Work: His Contribution to Musical Theory." *World Journal of Surgery* 21 (1997): 569–71.
McGann, Jerome. *The Poetics of Sensibility*. Oxford: Oxford University Press, 1996.
McGill, Anthony, et al. "Musicians on How to Bring Racial Equity to Auditions." *New York Times*, September 10, 2020. https://www.nytimes.com/2020/09/10/arts/music/diversity-orchestra-auditions.html.
McGuire, Charles Edward. *Music and Victorian Philanthropy: The Tonic Sol-Fa Movement*. Cambridge: Cambridge University Press, 2009.
McKeever, Natasha. "Can a Woman Rape a Man and Why Does It Matter?" *Criminal Law and Philosophy* 13 (2019): 599–619.

McQueen, Garrett. "The Power (and Complicity) of Classical Music." *Your Classical*, June 5, 2020. https://www.yourclassical.org/story/2020/06/05/the-power-and-complicity-of-classical-music.

———. "SNL's Public Indictment of Orchestra Culture." *Represent Classical*, April 23, 2022. https://representclassical.com/2022/04/23/snl-public-indictment-of-orchestra-culture/.

"Medicinal Music." *Musical Times and Singing Class Circular* 32, no. 584 (October 1, 1891): 587–88.

Melville, Herman. *Pierre: Or, The Ambiguities*. London: Penguin Books, 1996.

Mendes, Peter. *Clandestine Erotic Fiction in English, 1800–1930: A Bibliographical Study*. New York: Routledge, 1993.

Midgette, Anne. "How (Not) to Behave: Manners and the Classical Music Audience." *Washington Post*, April 29, 2015. https://www.washingtonpost.com/entertainment/music/how-not-to-behave-manners-and-the-classical-music-audience/2015/04/29/111a2aa4-bc72-11e4-b274-e5209a3bc9a9_story.html.

Miezkowski, Sylvia. *Resonant Alterities: Sound, Desire and Anxiety in Non-Realist Fiction*. New York: Columbia University Press, 2014.

Millgate, Michael. *Thomas Hardy's Library at Max Gate: Catalogue of an Attempted Reconstruction*. University of Toronto Library. https://hardy.library.utoronto.ca/.

"Miscellaneous Concerts." *Musical World* 65, no. 29 (June 16, 1887): 559.

Mitchell, Nicholas. "The Mystery of Music." In *London in Light and Darkness*, 233–35. London: William Tegg, 1871.

Molteno, Ada. "Ladies as Orchestral Players by One of Them." *Orchestral Association Gazette*, no. 6 (March 1894): 63–64.

moore, madison. *Fabulous: The Rise of the Beautiful Eccentric*. New Haven, CT: Yale University Press, 2018.

Moreno, Jairo. "Body'n'Soul? Voice and Movement in Keith Jarrett's Pianism." *Musical Quarterly* 83, no. 1 (Spring 1999): 75–92.

Morgan, Benjamin. *The Outward Mind: Materialist Aesthetics in Victorian Science and Literature*. Chicago: University of Chicago Press, 2017.

Morgan, Joan. "Why We Get Off: Moving Towards a Black Feminist Politics of Pleasure." In *Pleasure Activism: The Politics of Feeling Good*, edited by adrienne maree brown, 81–98. Chico, CA: AK Press, 2019.

Morrison, Richard. *Orchestra: The LSO; A Century of Triumph and Turbulence*. London: Faber & Faber, 2004.

Moss, Gemma. "Classical Music and Literature." In *Sound and Literature*, edited by Anna Snaith, 92. Cambridge: Cambridge University Press, 2020.

Mulock, Dinah. "The Last House in C—— Street." In *The Oxford Book of Victorian Ghost Stories*, edited by Michael Cox and R. A. Gilbert, 44–54. Oxford: Oxford University Press, 2003.

Muñoz, José Esteban. *Cruising Utopia: The Then and There of Queer Futurity*. New York: New York University Press, 2009.

"Music." *Illustrated London News* 69, no. 1931 (July 29, 1876): 114–15.

"Music as a Relief to Pain." *Musical Times and Singing-Class Circular* 22, no. 463 (September 1, 1881): 458.
"The Musical Artists' Society." *Musical Times* 34, no. 604 (June 1, 1893): 346.
Musical Times and Singing-Class Circular 35, no. 617 (July 1, 1894): 482.
Musical Times and Singing-Class Circular 37, no. 638 (April 1, 1896): 266.
"Musical Vibrations." *Weston-super-Mare Gazette, and General Advertiser* 41, no. 2210 (October 21, 1885): 3.
Nead, Lynda. *Victorian Babylon: People, Streets and Images in Nineteenth-Century London*. New Haven, CT: Yale University Press, 2005.
Neilson, Brett. "Hardy, Barbarism, and the Transformations of Modernity." In *Thomas Hardy and Contemporary Literary Studies*, edited by Tim Dolin and Peter Widdowson, 65–80. London: Palgrave Macmillan, 2004.
"Nerves of the Violin." *The Strad* 5, no. 57 (January 1895): 279.
Newman, Ernest. "The World of Music: A Physiology of Criticism." *Sunday Times*, December 16, 1928: 7.
Newington, H. Hayes. "Some Mental Aspects of Music." *Journal of Mental Science* 43 (October 1897): 704–23.
Nordau, Max. *Degeneration*. London: William Heinemann, 1895.
Nordlinger, Jay. "The Yuja 'n' Jaap Show." *New Criterion* (blog), March 5, 2018. https://newcriterion.com/blogs/dispatch/the-yuja-n-jaap-show.
Notices of the Proceedings at the Meetings of the Members of the Royal Institution of Great Britain. Vols. 7–9. London: William Clowes and Sons, 1875–81.
"Novello, Ewer, & Co.'s Music Primers" (advertisement). *Musical Times and Singing Class Circular* 21, no. 451 (September 1, 1880): 470.
O'Modhrain, Sile, and R. Brent Gillespie. "Once More, with Feeling: Revisiting the Role of Touch in Performer-Instrument Interaction." In *Musical Haptics*, edited by Stefano Papetti and Charalampos Saitis, 11–28. Berlin: Springer, 2018.
Offences Against the Person Act. 1861. In *Nineteenth-Century Writings on Homosexuality: A Sourcebook*, edited by Chris White, 42–43. New York: Routledge, 1999.
"On the Physiology of Pianoforte Playing, with a Practical Application of a New Theory." *Musical Standard* 32, no. 14 (January 1888): 18–19.
"The Organ World: On the Physiology of Pianoforte Playing, with a Practical Application of a New Theory." *Musical World* 68, no. 3 (November 3, 1888): 849–50.
Palus, Shannon. "How to Think About the Debate Over the Phrase 'Pregnant People.'" *Slate*, July 9, 2022. https://slate.com/technology/2022/07/pregnant-people-inclusive-language-gender-debate.html.
Parke, William T. *Musical Memoirs: Comprising an Account of the General State of Music in England, from the First Commemoration of Handel, in 1784, to the Year 1830*. London: Henry Colburn and Richard Bentley, 1830.
Parsons, Flora T. *Callisthenic Songs Illustrated: A New and Attractive Collection of Callisthenic Songs Beautifully Illustrated*. New York: Ivison, Blakeman, Taylor, 1869.

Peak, Anna. "The Condition of Music in Victorian Scholarship." *Victorian Literature and Culture* 44, no. 2 (2016): 423–37.
Peraino, Judith. "The Same, but Different: Sexuality and Musicology, Then and Now." *Journal of the American Musicological Society* 66, no. 3 (Fall 2013): 825–30.
Perkin, J. Russell. *Theology and the Victorian Novel*. Montreal: McGill University Press, 2009.
"Philharmonic Concerts." *Athenaeum: Journal of Literature, Science, and the Fine Arts* 347, no. 21 (June 1834): 475.
"Physiological Effects of Music." *Canterbury Journal and Farmer's Gazette*, no. 3,017 (May 26, 1894): 6.
Picker, John. *Victorian Soundscapes*. Oxford: Oxford University Press, 2003.
Pite, Ralph. *Thomas Hardy: The Guarded Life*. New Haven, CT: Yale University Press, 2007.
Pope, Rebecca. "The Diva Doesn't Die: George Eliot's 'Armgart.'" *Criticism* 32, no. 4 (Fall 1990): 469–83.
Poskett, James. *Materials of the Mind: Phrenology, Race, and the Global History of Science, 1815–1920*. Chicago: University of Chicago Press, 2019.
Powell, Michael. "A Vanishing Word in the Abortion Debate: 'Woman.'" *New York Times*, June 8, 2022. https://www.nytimes.com/2022/06/08/us/women-gender-aclu-abortion.html.
Priest, Eldritch. *Earworm and Event*. Durham, NC: Duke University Press, 2022.
"Prince Sprite." *Musical Times* 33, no. 593 (July 1, 1892): 441.
Prosser, Jay. *Second Skins: The Body Narratives of Transsexuality*. New York: Columbia University Press, 1998.
Przybylo, Ela. *Asexual Erotics: Intimate Readings of Compulsory Sexuality*. Columbus: Ohio State University Press, 2019.
Puar, Jasbir. *Terrorist Assemblages: Homonationalism in Queer Times*. Durham, NC: Duke University Press, 2007.
Pulham, Patricia. *Art and the Transitional Object in Vernon Lee's Supernatural Tales*. Aldershot: Ashgate, 2008.
Pykett, Lyn. "The Material Turn in Victorian Studies." *Literature Compass* 1, no. 1 (2004): 1–5.
Radau, Rodolphe. *Wonders of Acoustics: Or, The Phenomena of Sound*. New York: Charles Scribner's Sons, 1886.
Rebelo, Pedro. "Haptic Sensation and Instrumental Transgression." *Contemporary Music Review* 25, no. 1/2 (February/April 2006): 27–35.
"Review: *Physiological Aesthetics* by Grant Allen." *Popular Science Monthly* 11 (October 1877): 760.
"Review: *The Violin-Player* by Bertha Thomas." *The Graphic*, July 31, 1880: 119.
Richardson, Angelique. *Love and Eugenics in the Late Nineteenth Century: Rational Reproduction and the New Woman*. Oxford: Oxford University Press, 2003.
Riddell, Fraser. *Music and the Queer Body in English Literature at the Fin de Siècle*. Cambridge: Cambridge University Press, 2022.

Roberts, Maddy Shaw. "The 9-Year-Old Boy Who Blurted Out 'Wow!' at the End of a Concert Has Been Found." *Classic FM*, May 10, 2019. https://www.classicfm.com/music-news/orchestra-seeks-wow-child-concert/.

Rogers, Jillian C. *Resonant Recoveries: French Music and Trauma Between the World Wars*. Oxford: Oxford University Press, 2021.

Rosenman, Ellen Bayuk. *Unauthorized Pleasures*. Ithaca, NY: Cornell University Press, 2003.

Ross, Alex. "Hold Your Applause: Inventing and Reinventing the Classical Concert." Lecture at the Royal Philharmonic Society, London. *The Rest Is Noise* (blog), March 8, 2010. https://www.therestisnoise.com/2005/02/18/.

Roy, Deboleena. "Somatic Matters: Becoming Molecular in Molecular Biology." *Rhizomes*, no. 14 (Summer 2007). http://www.rhizomes.net/issue14/roy/roy.html.

"Royal Academy of Music." *Musical Times* 17, no. 390 (August 1, 1875): 170–71.

"Royal Academy of Music: Female Department." *Musical Times* 18, no. 414 (August 1, 1877): 390.

Rudy, Jason. *Electric Meters: Victorian Physiological Poetics*. Athens: Ohio University Press, 2009.

Russell, Tilden. "The Development of the Cello Endpin." *Imago Musicae* 4 (1985): 335–56.

———. "Endpin." *Grove Music Online*, Oxford University Press, 2001. https://doi.org/10.1093/gmo/9781561592630.article.08788.

Ryan, Vanessa. *Thinking Without Thinking in the Victorian Novel*. Baltimore: Johns Hopkins University Press, 2012.

Sacks, Oliver. *Musicophilia: Tales of Music and the Brain*. Rev. and expanded ed. New York: Vintage, 2007.

Saint-George, Henry. "The Bow, Its History, Manufacture and Use." *The Strad* 7, no. 78 (October 1896): 175.

Sanden, Paul. "Hearing Glenn Gould's Body: Corporeal Liveness in Recorded Music." *Current Musicology* 88 (Fall 2009): 7–34.

Sanders, Valerie. *Eve's Renegade: Victorian Anti-Feminist Women Novelists*. New York: Palgrave Macmillan, 1996.

Saturday Night Live. "Orchestra—SNL." YouTube video, April 16, 2022, 4:43. https://www.youtube.com/watch?v=KhctLo_qS10.

Scarry, Elaine. "Work and the Body in Hardy and Other Nineteenth-Century Novelists." *Representations* 3 (Summer 1983): 90–123.

Schaffer, Talia. "Feminism and the Canon." In *The Routledge Companion to Victorian Literature*, edited by Dennis Denisoff and Talia Schaffer, 273–83. New York: Routledge, 2020.

———. *Romance's Rival: Familiar Marriage in Victorian Fiction*. Oxford: Oxford University Press, 2016.

Scharwenka, Xaver. "The Octave Staccato." In *The Music of the Modern World, Illustrated in the Lives and Works of the Greatest Modern Musicians and in Reproduction of Famous Paintings, Etc.*, edited by Anton Seidl et al., 5–7. Boston: D. Appleton, 1895.

Schmidt, Leigh Eric. *Hearing Things: Religion, Illusion and the American Enlightenment*. Cambridge, MA: Harvard University Press, 2000.

Schuller, Kyla. *The Biopolitics of Feeling: Race, Sex, and Science in the Nineteenth Century*. Durham, NC: Duke University Press, 2018.

Scott, Derek B. *From the Erotic to the Demonic: On Critical Musicology*. Oxford: Oxford University Press, 2003.

———. *The Singing Bourgeois: Songs of the Victorian Drawing Room and Parlour*. Aldershot: Ashgate, 2001.

Scott, Marion. "British Women as Instrumentalists." *Music Student* 10, no. 9 (May 1913): 337–38.

———. *Violin Verses*. London: Walter Scott, 1905.

Seashore, Carl. "The Measurement of Musical Talent." *Musical Quarterly* 1, no. 1 (1915): 129–48.

Sedgman, Kirsty. *The Reasonable Audience: Theatre Etiquette, Behaviour Policing, and the Live Performance Experience*. London: Palgrave Macmillan, 2018.

Sedgwick, Eve Kosofsky. *Epistemology of the Closet*. Berkeley: University of California Press, 1990.

———. *Touching Feeling: Affect, Pedagogy, Performativity*. Durham, NC: Duke University Press, 2003.

Senf, Carol. Introduction to *The Heavenly Twins* by Sarah Grand, vii–xxxvii. Ann Arbor: University of Michigan Press, 1994.

"Sense of Absolute Pitch." *Musical Herald*, no. 577 (April 1, 1896): 108–9.

Showalter, Elaine. *Sexual Anarchy: Gender and Culture at the Fin de Siècle*. New York: Viking, 1990.

Simpson, Hannah. "Tics in the Theatre: The Quiet Audience, the Relaxed Performance, and the Neurodivergent Spectator." *Theatre Topics* 28, no. 3 (2018): 227–38.

The Sins of the Cities of the Plain. Edited by Wolfram Setz. Richmond, VA: Valancourt Classics, 2012.

Solie, Ruth. "No 'Land without Music' After All." *Victorian Literature and Culture* 32, no. 1 (2004): 261–76.

"Some Physiological Effects of Music." *Stroud Journal* 39, no. 2056 (October 20, 1893): 8.

Spampinato, Erin. "Rereading Rape in the Critical Canon: Adjudicative Criticism and the Capacious Conception of Rape." *differences* 32, no. 2 (2021): 122–60.

Spencer, Herbert. "The Origin and Function of Music." In *Essays: Scientific, Political, and Speculative*, vol. 2. London: Williams and Norgate, 1891.

———. *The Principles of Psychology*. 2nd ed. London: Williams and Norgate, 1870.

Steege, Benjamin. *Helmholtz and the Modern Listener*. Cambridge: Cambridge University Press, 2012.

Sterne, Jonathan. *The Audible Past: Cultural Origins of Sound Reproduction*. Durham, NC: Duke University Press, 2003.

Stone, W. H. *The Scientific Basis of Music*. London: Novello, Ewer, 1878.

Stryker, Susan. "Transgender Studies: Queer Theory's Evil Twin." In *Feminist and Queer Theory: An Intersectional and Transnational Reader*, edited by L. Saraswati and Barbara L. Shaw, 70–72. Oxford: Oxford University Press, 2021.

Sully, James. *Sensation and Intuition: Studies in Psychology and Aesthetics*. London: Henry S. King, 1874.

Sutcliffe, Wallace. "Ladies as Orchestral Players." *Orchestral Association Gazette* 5 (February 1894): 48–49.

Sutton, Emma. "'The Music Spoke for Us': Music and Sexuality in *Fin-de-siècle* Poetry." In *The Figure of Music in Nineteenth-Century British Poetry*, edited by Phyllis Weliver, 213–29. Aldershot: Ashgate, 2005.

Tallbear, Kim. "An Indigenous Reflection on Working Beyond the Human/Not Human." *GLQ* 21, nos. 2–3 (June 2015): 230–35.

Tanaka, Atau, and Marco Donnarumma. "The Body as Musical Instrument." In *The Oxford Handbook of Music and the Body*, edited by Youn Kim and Sander L. Gilman, 79–96. Oxford: Oxford University Press, 2019.

Taylor, Clare. *Women, Writing, and Fetishism, 1890–1950: Female Cross-Gendering*. Oxford: Oxford University Press, 2003.

Taylor, Jodi. "Sound Desires: Auralism, the Sexual Fetishization of Music." In *The Oxford Handbook of Music and Queerness*, edited by Fred Everett Maus and Sheila Whiteley, 277–94. Oxford: Oxford University Press, 2022.

———. "Taking It in the Ear: On Music-Sexual Synergies and the (Queer) Possibility That Music Is Sex." *Continuum* 26, no. 4 (August 2012): 603–14.

Taylor, Sedley. *Sound and Music: A Non-Mathematical Treatise on the Physical Constitution of Musical Sounds and Harmony, Including the Chief Acoustical Discoveries of Professor Helmholtz*. London: Macmillan, 1883.

Teleny: Étude Physiologique. Translated by Charles Hirsch. Paris: La Musardine, 2009.

Teleny, or the Reverse of the Medal. New York: Mondial, 2006.

Thierauf, Doreen. "*Daniel Deronda*, Marital Rape, and the End of Reproduction." *Victorian Review* 43, no. 2 (Fall 2017): 247–69.

———. "The Hidden Abortion Plot in George Eliot's *Middlemarch*." *Victorian Studies* 56, no. 3 (Spring 2014): 479–89.

———. "Tending to Old Stories: *Daniel Deronda* and Hysteria, Revisited." *Victorian Literature and Culture* 46, no. 2 (June 2018): 443–65.

Thomas, Bertha. *The Violin-Player*. London: Richard Bentley and Son, 1880.

Thomas, Kate. *Postal Pleasures*. Oxford: Oxford University Press, 2012.

Thurschwell, Pamela. *Literature, Technology, and Magical Thinking, 1880–1920*. Cambridge: Cambridge University Press, 2001.

Tod, David. *The Anatomy and Physiology of the Organ of Hearing*. London: Longman, Rees, Orme, Brown, Green, and Longman, 1832.

Tommasini, Anthony. "To Make Orchestras More Diverse, End Blind Auditions." *New York Times*, July 16, 2020. https://www.nytimes.com/2020/07/16/arts/music/blind-auditions-orchestras-race.html.

Toop, David. *Sinister Resonance: The Mediumship of the Listener*. London: Bloomsbury, 2010.
Towheed, Shafquat. "The Science of Musical Memory: Vernon Lee and the Remembrance of Sounds Past." In *Words and Notes in the Long Nineteenth Century*, edited by Phyllis Weliver and Katharine Ellis, 104–22. Suffolk, UK: Boydell and Brewer, 2013.
Tracy, James R. "The Power of Music Over Animals." *Musical Standard*, August 11, 1894: 106–7.
TreasuresDelightsEtc. "Music Teacher Notebook." Etsy. Accessed September 9, 2023. https://www.etsy.com/sg-en/listing/211797339/notebook-music-is-my-escape-music.
Trevelyan, Janet. *The Life of Mrs. Humphry Ward*. New York: Dodd, Mead, 1923.
Trevenen Dawson, E. M. "How to Behave at Concerts." *Monthly Musical Record* 28, no. 326 (February 1, 1898): 28–29.
Trippett, David. *Wagner's Melodies: Aesthetics and Materialism in German Musical Identity*. Cambridge: Cambridge University Press, 2013.
Trollope, Anthony. *The Letters of Anthony Trollope, 1851–1882*. Edited by Bradford Allen Booth. Oxford: Oxford University Press, 1951.
———. *The Warden*. Edited by Nicholas Shrimpton. Oxford: Oxford University Press, 2009.
Trower, Shelley. *Senses of Vibration: A History of the Pleasure and Pain of Sound*. London: Continuum, 2012.
Tyndall, John. *Sound*. 1st ed. London: Longmans, Green, 1867.
———. *Sound*. 3rd ed. New York: D. Appleton, 1898.
Upton, George. *Musical Memories*. Chicago: A. C. McClurg, 1908.
van der Straeten, Edmund S. J. *The Technics of Violoncello Playing*. London: The Strad, 1898.
———. "The Technics of Violoncello Playing." *The Strad* 6, no. 61 (May 1895): 18–20.
Vicinus, Martha. "Turn-of-the-Century Male Impersonation: Rewriting the Romance Plot." In *Sexualities in Victorian Britain*, edited by Andrew H. Miller and James Eli Adams, 187–213. Bloomington: Indiana University Press, 1996.
Vogel, Stephan. "Sensations of Tone, Perception of Sound, and Empiricism: Helmholtz's Physiological Aesthetics." In *Hermann von Helmholtz and the Foundations of Nineteenth-Century Science*, edited by David Cahan, 259–90. Berkeley: University of California Press, 1993.
"Voice Failure and Its Attendant Ailments." *Musical Herald*, no. 619 (October 1, 1899): 309.
Ward, Mary Augusta. *The Case of Richard Meynell*. Garden City, NY: Doubleday, 1911.
———. *Robert Elsmere*. Edited by Miriam Elizabeth Burstein. Brighton, UK: Victorian Secrets, 2013.

Warner, Michael. *The Trouble with Normal: Sex, Politics, and the Ethics of Queer Life*. Cambridge, MA: Harvard University Press, 1999.
Weber, William G. *Music and the Middle Class: The Social Structure of Concert Life in London, Paris, and Vienna between 1830 and 1848*. New York: Holmes and Meier, 1975.
Weliver, Phyllis. Introduction to *The Figure of Music in Nineteenth-Century British Poetry*, edited by Phyllis Weliver, 1–24. Aldershot: Ashgate, 2005.
———. *The Musical Crowd in English Fiction: Class, Culture, and Nation*. London: Palgrave Macmillan, 2006.
———. "A Score of Change: Twenty Years of Critical Musicology and Victorian Literature." *Literature Compass* 8, no. 10 (2011): 776–94.
———. *Women Musicians in Victorian Fiction, 1860–1900: Representations of Music, Science and Gender in the Leisured Home*. Aldershot: Ashgate, 2000.
Wellings, Christopher. "Dangerous Desires: The Uses of Women in *Teleny*." *Oscholars: Special* Teleny *Issue* (Autumn 2008). http://www.oscholars.com/Teleny/wellings.htm.
West, Anna. *Thomas Hardy and Animals*. Cambridge: Cambridge University Press, 2017.
Weston, Pamela. *Clarinet Virtuosi of the Past*. London: Robert Hale, 1971.
Wiegman, Robyn, and Elizabeth A. Wilson. "Introduction: Antinormativity's Queer Conventions." *differences* 26, no. 1 (2015): 1–25.
Wilbourne, Emily. "The Queer History of the Castrato." In *The Oxford Handbook of Music and Queerness*, edited by Fred Everett Maus and Sheila Whiteley, 441–54. Oxford: Oxford University Press, 2022.
Wilde, Oscar. *De Profundis*. London: Methuen, 1912.
———. *The Picture of Dorian Gray*. Oxford: Oxford World's Classics, 2008.
Wilde, William Robert. *Practical Observations on Aural Surgery and the Nature and Treatment of Diseases of the Ear*. Philadelphia: Blanchard & Lea, 1853.
Willey, Angela. "Biopossibility: A Queer Feminist Materialist Science Studies Manifesto, with Special Reference to the Question of Monogamous Behavior." In *Feminist and Queer Theory: An Intersectional and Transnational Reader*, edited by L. Saraswati and Barbara Shaw, 509–22. Oxford: Oxford University Press, 2021.
———. *Undoing Monogamy: The Politics of Science and the Possibilities of Biology*. Durham, NC: Duke University Press, 2016.
Wills, Matthew. "Mary Somerville, Queen of 19th Century Science." *JSTOR Daily*, March 2, 2016. https://daily.jstor.org/mary-somerville-queen-of-19th-century-science/.
Wilson, Elizabeth A. *Gut Feminism*. Durham, NC: Duke University Press, 2015.
———. *Psychosomatic: Feminism and the Neurological Body*. Durham, NC: Duke University Press, 2004.
Wilson, Mary Katherine, Sarah Marczynski, and Elizabeth O'Brien. "Ethical Behavior of the Classical Music Audience." *Ethical Human Psychology and Psychiatry* 16, no. 2 (2014): 120–26.

Wilt, Judith. "The Romance of Faith: Mary Ward's *Robert Elsmere* and *Richard Meynell.*" *Literature and Theology* 10, no. 1 (March 1996): 33–43.

Winn, Edith Lynwood. "The Study of the Violin for Girls." *Musical World* 1, no. 10 (November 1901): 132.

Wollaston, William Hyde. "On Sounds Inaudible by Certain Ears." Proceedings of the Royal Society of London, Philosophical Transactions of the Royal Society, 1820.

Wolfreys, Julian. *Victorian Hauntings: Spectrality, Gothic, the Uncanny and Literature.* London: Palgrave Macmillan, 2002.

"Women in Orchestras." *Musical Standard* 46 (January 20, 1894): 47.

Wood, Elizabeth. "Sapphonics." In *Queering the Pitch*, edited by Philip Brett and Elizabeth Wood, 27–66. New York: Routledge, 1994.

Woods, Livia Arndal. "Now You See It: Concealing and Revealing Pregnant Bodies in *Wuthering Heights* and *The Clever Woman of the Family.*" *Victorian Network* 6, no. 1 (Summer 2015): 32–54.

Young, G. M. Introduction to *The Nebuly Coat and The Lost Stradivarius*, by John Meade Falkner, vii–x. Oxford: Oxford University Press, 1954.

Zevin, Gabrielle. *Tomorrow and Tomorrow and Tomorrow.* New York: Knopf, 2022.

Zon, Bennett. *Evolution and Victorian Musical Culture.* Cambridge: Cambridge University Press, 2017.

———. *Music and Metaphor in Nineteenth-Century British Musicology.* Aldershot: Ashgate, 2000.

Index

Ablow, Rachel, 114, 224n33
absolute music, 38, 209n74
acoustic chair, 32, *33*, 34–35
acoustical instruments, 32, *33*
acoustical science, 2–3, 5–6; aesthetic debates and, 37–42; class and, 7–8, 41–42, 45–46; in education, 44–46; medicine and, 44–45; race and, 7–8, 41–42, 45–46. *See also* physical acoustics; physiological acoustics
Adorno, Theodor, 13
The Adventures of a Schoolboy (anonymous), 130
aeolian harp, 31, 87, 207n30
Ahmed, Sara, 4–5, 151, 153, 159, 166
Aldridge, Amanda, 8
Allen, Grant, 39, 112–13
Amin, Kadji, 8
animals (and music): musical response of, 37, 41, 96; sexual selection and, 31
Anolik, Ruth Bienstock, 173
anonymous auditions, 192–93, 197, 242n18
Anstruther-Thomson, Kit, 39
antibiologism, 12, 77–78
Arnim, Anna Leffler, 67, *68*
asexuality, 3, 16, 82–83, 151
Asquith, Mark, 116, 124
Auerbach, Emily, 53
aural sex, 17; musical rape and, 104–5; in *Teleny*, 133, 141–43, 231n69. *See also* Taylor, Jodi

Auyoung, Elaine, 164

Bagby, Albert Morris, 54
Baitz, Dana, 14, 83
Baker, Frederick Charles, 113
Barbell or Wand Exercises (McCartney), 45–46
Barnaby, Joseph, 61
Barnard, Charles, 61–62
Bashant, Wendy, 86
Baughan, Edward, 44
Baumgardner, Astrid, 196
von Bayros, Franz, 130, *131*
Beardsley, Aubrey, 130
The Beautiful (Lee), 40
Beer, Gillian, 9
Beethoven, 1–2, 5, 38
Behar, Katherine, 151
Behnke, Emil, 43
Bennati, Francesco, 67
Bennett, Jane, 150
Besant, Walter, 54
Best, Stephen, 173
Billroth, Theodor, 64
Blundell, Mary Elizabeth. *See* Francis, M. E.
Bosanquet, R. H. M., 93
Braddon, Mary Elizabeth, 53
Bradley, Katherine. *See* Field, Michael
Braidotti, Rosi, 150
Brain, Robert Michael, 39
Brett, Philip, 127, 129
Brewster, David, 28

Bristow, Joseph, 131, 133
Broadhouse, John, 44
Browne, Lennox, 43
Buggery Act, 205n76
Bumke, Oswald, 72
Burdett, Carolyn, 11–12, 39, 108
Burgan, Mary, 15, 18
Burke, Edmund, 31
Buse, Peter, 170
Butler, Judith, 80, 82, 186, 219n9

Caballero, Carlo, 181–182
Callisthenic Songs Illustrated (Parsons), 46, *46*
Campbell, Daisy Rhodes, 54, 56
carnal musicology, 14, 83
Carpenter, Edward, 129
The Case of Richard Meynell (Ward), 69
Castle, Terry, 172
castrato: as erotic figure, 171; as otherworldly, 240n75; in queer musicology, 171, 237n13; sexual transgression and, 128, 171, 182, 237n13; in "A Wicked Voice," 180–186
Chant, L. Ormiston, 45–46
Chatterjee, Ronjaunee, 7
Cheng, William, 102, 127, 192–93, 242n18
Chladni, Ernst, 27, *28*
Chomet, Hector, 36, 139, 183
Chow, Jeremy, 148
Christoff, Alicia Mireles, 7
Clapp-Itnyre, Alisa, 15
Clarissa (Richardson), 104
classical music (21st century): anonymous auditions in, 192–93, 242n18; body erasure in, 189–98; concert etiquette in, 191–92; COVID-19 and, 194; disability and, 194–96; health consequences in, 194; race and, 20–21, 193; terminology of, 201n26. *See also* Western classical music
Cohen, Cathy, 4
Cohen, Ed, 133

Cohen, William, 10, 13
Cohn, Elisha, 111
Colby, Vineta, 181
Coleridge-Taylor, Samuel, 8
A Complete Course of Wrist and Finger Gymnastics (Arnim), 67, *68*
concert dress: "concert black" in, 20, 192; SNL skit about, 189–90
concert etiquette, 7, 38, 190; classist and racist ideologies in, 193–95; on coughing, 191–92; disability and, 194–96; manuals for, 38; relaxed performances and, 195–96
concert hall: curation of, 38, 209n77; eroticism of, 1–5; as haunted, 186–87; orgasm in (*Teleny*), 127–30, 134–36; queering of, 14–15, 19
Conrad, Joseph, 54, 166
Cooper, Edith. *See* Field, Michael
Corti, Alfonso, 34
COVID-19 pandemic, 194
Cox, Kimberly, 104, 165
Cox, Michael, 170, 173
cross-dressing, 18, 79–98. *See also* gender performativity; *The Heavenly Twins*; *The Violin Player*
Curtis, John Harrison, 32, *33*
Curwen, John, 46
Cusick, Suzanne: on "lesbian" relationship with music, 127; on music as sex, 4, 14, 17, 127, 141, 143, 151; on music torture, 102

da Sousa Correa, Delia, 53, 107–108, 159–160
dactylergon, 154
Dalley, Lana, 103, 108
Daly, Nicholas, 174–175, 239n43
Dames, Nicholas, 10, 39, 210n86, 211n101
dance music, 112–13
Daniel Deronda (Eliot), 53, 104, 107
Darwin, Charles, 31, 90, 115, 129, 207n35. See also *The Descent of Man*
Davies, James Q., 155, 240n75

Index | 273

Dawson, E. M. Trevenen, 190–91
deafness. *See* hearing impairment
Dean, Tim, 3, 143
deidealization, 8
Delgado, Maria, 152–53
The Descent of Man (Darwin), 31, 59
Desperate Remedies (Hardy), 19; musical rape in, 102, 104, 115–20
Deutsch, David, 130
Dickens, Charles, 9, 147, 166–67, 172; *Great Expectations* by, 232n6; human-instrument intimacies of, 147–52. See also *Dombey and Son*
Dinshaw, Carolyn, 173
disability and classical music, 194–96
Dixon, J. Herbert, 60
Dogiel, Alexandre, 36, 183
Dombey and Son (Dickens), 19–20; cello in, 147–49
"Dr. Woolacott" (Forster), 169
Du Maurier, George, 57, 58; *Trilby* by, 104
The Duenna of a Genius (Francis), 17, 18; acoustical science and, 52, 69–78; gender politics in, 54, 62, 69–78; as "lady violinist" novel, 52, 62; marriage in, 77; metaphysical language in, 75–76; muscle memory in, 73–74; romance in, 74–77
Dworkin, Afa, 197

ear: anatomy of, 32, 34, 36; as erogenous zone, 141, 143
earworm: as music torture, 137; sexual desire and, 136–37, 140; for sound preservation, 138; in *Teleny*, 136–38, 140; theorization of, 213n45; as universal, 137, 140
Edgers, Geoff, 192
Edison, Thomas, 166
education: musical degrees in, 44, 212n127–29; musical drills in, 45–46; women, music and, 55–56
Eidsheim, Nina, 12–13
Eliot, George, 5–6, 9; *Daniel Deronda* by, 53, 104, 107; on divas, 53–54; Helmholtz and, 107; human-instrument intimacies in, 147–52; "Mr. Gilfil's Love Story" by, 107; musical interests of, 107–8, 224n33; *Romola* by, 101; "Stradivarius" by, 157. See also *The Mill on the Floss*
Ellis, Alexander, 29, 36
Ellis, Havelock, 37, 41–42, 103, 129, 139
embodied music theory. *See* carnal musicology
English Musical Renaissance, 2
the erotic (theories of), 3–4, 15–17, 200n15
erotohistoriography, 20

"The Fair Sex-Tett" (Du Maurier), 57, 58
Falkner, John Meade, 5, 174. See also *The Lost Stradivarius*
Far from the Madding Crowd (Hardy), 163–64
Faraday, Michael, 28
female violinists. *See* lady violinist
feminist musicology, 3–4, 7, 17; acoustical science and, 11–12
feminist theory, 6–10, 11–12; the body in, 11–12, 77–78; the erotic in, 3–4, 15–17, 200n15; posthumanism in, 150, 233n18; and Victorian studies, 16–17, 200n24
Fennell, James, 41
Fétis, François-Joseph, 40
"The Fiddler of the Reels" (Hardy), 19; musical rape in, 102, 104, 115–25
Fidler, Florence G., 61, 98
Field, Michael, 129
Filipowicz, Elise, 59
Finlay, Mildred, 54
Forster, E. M., 166, 169
Fothergill, Jessie, 54
Foucault, Michel, 11, 143
Francis, M. E., 16–17, 52; as devotee of music, 70. See also *The Duenna of a Genius*

Frank, C. E., 42
Freccero, Carla, 172
Freeman, Elizabeth, 4, 20, 173
Freitas, Roger, 171
Friedman, Dustin, 12, 16, 133–34
Fuss, Diana, 172

Gall, François Joseph, 64
Galton, Francis, 90
Garcia, Manuel, 43–44
Gautherot, Madame Louisa, 59
gender performativity, 18, 77, 80, 82–83, 88; musical performance and, 82–83. *See also* cross-dressing
Gender Trouble (Butler), 82, 219n9
George, W. Tyacke, 56, 60–61, 73
ghost stories, 19; Gothic literature *vs.*, 170; popularity of, 170; queerness of, 172–73. *See also* musical hauntings
Gilbert, R. A., 170, 173
Gillett, Paula, 53, 59, 76, 215n57, 217n97
Gissing, George, 54
Glover, Robin Bickerstaff, 191
Goddard, Joseph, 38
Godfrey, Elizabeth, 54
Gothic literature, 170, 173
Gould, Glenn, 152
gramophone, 138, 166
Grand, Sarah, 5–6, 12, 82; and eugenics, 12, 82, 90; musical heroines of, 84; in New Woman movement, 85. *See also The Heavenly Twins*
Gray, Beryl, 108, 114
Great Expectations (Dickens), 232n6
Green, Lucy, 56
Grétry, André, 36
Grosz, Elizabeth, 12, 78
Grove, George, 115
Guild of St. Cecilia, 45
Gurman, Elissa, 103
Gurney, Edmund, 36, 41, 116, 129, 137

Haefele-Thomas, Ardele, 16
Hager, Lisa, 16
Halberstam, Jack, 4
Hall, Marie, 55
Halperin, David, 143
Handel, George Friedrich, 232n6
Hanslick, Eduard, 40, 42
Hardy, Thomas, 5–6, 13, 101, 167; *Far from the Madding Crowd* by, 163–64; human-instrument intimacies of, 147–52; *Jude the Obscure* by, 164; musical interests of, 115–16; on physiological science, 115–16; *Tess of the D'Urbervilles* by, 104, 117, 223n18; *Under the Greenwood Tree* by, 163. *See also Desperate Remedies*; "The Fiddler of the Reels"; "Haunting Fingers"
Harford, Frederick Kill, 45
Harris, T. F., 44
Hartley, David, 31, 207n31, 218n119
"Haunting Fingers" (Hardy), 20, 149; instrument perspective in, 163–66; preservation and memory in, 165–66
Haweis, H. R., 62, 175, 239n41
Heap, Swinderton, 43
hearing impairment, 32, 36, 37
The Heavenly Twins (Grand), 18; Book I in, 90; Book IV in, 84–88, 220n16; Book V in, 88–89; cross-dressing female violinist in, 80, 84–90; gender transformation process in, 80–81; metaphysical terminology in, 87
Helmholtz, Hermann von, 2–3, 9–10, 12; acoustical experiments by, 28–29, 32–35; on compound and partial tones, 177–78; Eliot and, 107; on material ear, 34, 208n48; on music *vs.* noise, 29; *On the Sensations of Tone as a Physiological Basis for the Theory of Music* by, 10, 28–29, 34–36; optics studies by, 28; physiological studies by, 32, 34–36;

Index | 275

popularity of, 43; on pure tones, 93–94; resonator invented by, 34, 35; on sympathetic vibration, 29, 34–36; Tyndall and, 29. See also sympathetic vibration
Henchman, Anna, 232n14
Henry, Nancy, 107
heroine at the piano, 15, 18, 51, 53
Herschel, John, 28
"Higher Harmonies" (Lee), 39
Hirsch, Charles, 140
Hirschfeld, Magnus, 129
Hoffmann, E. T. A., 38, 209n73
Holst, Gustav, 115
Holzer, Kellie, 103, 108
homosexuality: as crime, 19, 134, 205n76; music associations with, 12, 128–29, 228n4; music-as-metaphor for, 13, 129–31, 174, 203n54; as natural, 132–34, 230n24; in pornography, 128–30, 228n3; in "Sinfonia Eroica" (Levy), 1–3, 13; *Teleny* defense of, 133–41; Victorian sexology on, 12, 15, 183
Honeyman, W. C., 60, 80
Hopkins, Gerard Manley, 9
"How to Behave at Concerts" (Dawson), 190–91
Howard, Lady Mabel, 54
Hughes, John, 116
Hughes, Linda K., 199n7
human-instrument intimacy, 19–20; in acoustical science, 152–57; of Dickens, 147–52; of Eliot, 147–52, 157–60; of Hardy, 147–52, 163–66; queer kinship and, 151, 156–57; in queer musicology, 152; self-playing instruments and, 166–67; sympathetic kinship and, 157–60; of Trollope, 147–52, 160–63. See also instruments
Hunter, J. Ewing, 45
Hush! (Tissot), 55
Huxley, T. H., 41, 62, 115, 210n86, 210n91

Igudesman and Joo, 196
The Influence of Music on Health and Life (Chomet), 36
instruments: agency of, 149–50, 163–166, 232n14, 233n17; as anthropomorphized, 155–56, 163–66; bodies of, 152, 156–57, 165; as erotic, 147–67; exercises for, 153–55, *154*; poems from perspective of, 164, 236n76; as self-playing, 166–67
inversion, 128–129, 132, 228n5. See also homosexuality

Jarvis, Claire, 13
Joachim, Joseph, 59
"Joachim's Rival," 59, *59*
Johnson, Barbara, 105
Johnson, Jenny Olivia, 223n4
Johnson, Julian, 201n26
Jones, F. Leslie, 46
Joy, Eileen, 151
Joyce, James, 166
Joyce, Simon, 16
Jude the Obscure (Hardy), 164

Kahan, Benjamin, 133
Keen, Suzanne, 116
Keller, Helen, 32
Kettle, Martin, 191
Killeen, Jarlath, 162
Koenig, Rudolph, 34
Krohn-Grimberghe, Lukas, 194–95

Labouchère Amendment, 134
ladies' orchestras, 56, 91, 214n21
Lady Audley's Secret (Braddon), 53
lady violinist, 6, 11, 16, 18; acoustical science and, 52–53; advocates for, 61–62; backlash against, 55–61; education for, 55, *55*; moral corruption of, 56–57; in orchestras, 56, 91, 98, 214n21; as siren or diva, 53–54, 70; strength of, 61–62; as ugly, 57, 59; in Victorian England,

lady violinist *(continued)*
55–62; violin as devil's instrument and, 57. See also *The Duenna of a Genius*; *Robert Elsmere*
Lang, Anna, 61
Latour, Bruno, 233n17
Law, Joe, 13, 129, 229n16
Lazarus, Henry, 91
Le Guin, Elisabeth, 187
Lee, Vernon, 5; *The Beautiful* by, 40; "Higher Harmonies" by, 39; *Music and Its Lovers* by, 40, 181; on physiological aesthetics, 39–40, 181, 211n102; "Winthrop's Adventure" by, 240n90. See also "A Wicked Voice"
Leopold, Amadéus, 193–94, 197
Levy, Amy. See "Sinfonia Eroica"
Lewes, George Henry, 73, 107, 115–116, 159, 162
Lindsay, Lady Blanche, 56–57, 61, 67
Lizzo, 189
Lorde, Audre, 3, 9, 17, 52, 151, 200n15
The Lost Stradivarius (Falkner), 20; acoustical science in, 169–80
Lubey, Kathleen, 104, 229n15
Lunn, Henry C., 45, 60, 73

MacCormack, Patricia, 150
Mack, Louise, 54
MacKinnon, Catharine, 105
Maddocks, Fiona, 191–92
Mahler, Gustav, 38, 187
Mahoney, Kristin, 8, 16
"The Maker of Violins" (anonymous), 148
Manne, Kate, 106, 226n74
Marchesi, Mathilde, 43–44
Marcus, Sharon, 16, 173
Marsh, Richard, 169
Marshall, Florence Ashton, 91
Martín Alcoff, Linda, 103, 104, 105
Maus, Fred Everett, 17
Maxwell, Catherine, 181
McCartney, R. H., 45–46

McGill, Anthony, 193, 197
McGuire, Charles, 46
McKendrick, J. G., 93
McQueen, Garrett, 190
medicine (and music), 44–45
Melville, Herman, 148
Midgette, Anne, 191
Miezkowski, Sylvia, 173, 181
The Mill on the Floss (Eliot), 19–20; dance scene in, 112–113; hand-grabbing in, 114; human-instrument encounters in, 149, 157–60; muscle memory in, 159; musical rape in, 102, 104, 107–15; sympathetic kinship in, 157–60
Millwood, Sasha Valeri, 192–93
Mitchell, Nicholas, 38
Molteno, Ada, 56, 61
moore, madison, 197
Morgan, Benjamin, 10, 116
Morgan, Joan, 200n15
Morrison, Richard, 56
"Mr. Gilfil's Love Story" (Eliot), 107
Müller, Johannes, 34
Muñoz, José Esteban, 8, 172
muscle memory (musical), 218n119; in *The Duenna of a Genius*, 73–74; in *The Mill on the Floss*, 159; in *Teleny*, 137–38; in *The Warden*, 162–63
music: as abstract ideal, 37–38, 42–43; as erotic, 4, 7, 14, 17, 104–5, 127, 133, 141, 143, 151; harmful potential of, 101–25, 223n4; inborn talent for, 64, 92; noise *vs.*, 29; sound *vs.*, 2
Music and Its Lovers (Lee), 40, 181
music technology, 166–67, 236n91
music therapy, 44, 45
music torture: in *Desperate Remedies*, 117–18; earworms as, 137; in "The Fiddler of the Reels," 123; in *The Heavenly Twins*, 89; physical traumas of, 102; in *Romola*, 101
musical drills, 45–46
musical hauntings, 20; acoustical science and, 174–80; concert

hall and, 186–87; in *The Lost Stradivarius* (Falkner), 169–80; in "A Wicked Voice" (Lee), 180–86. *See also* ghost stories
musical prodigies, 64, 79, 92
musical rape, 19, 102–3; aural sex and, 104–5; use of *rape* in, 102–3; victim blaming and, 114, 124. *See also* Eliot, George; Hardy, Thomas
music-as-metaphor: for homosexuality, 13, 129–30, 174; instruments and, 34–35, 148–150; in music-literature studies, 12–14; for sex, 4, 10, 13, 175
music-literature studies, 8, 12–14
"The Mystery of Music" (Mitchell), 38

"Nerves of the Violin," 156, 235n48
New Woman movement, 11, 85
New York Philharmonic, 193
Newington, H. Hayes, 73
Newman, Ernest, 41
Nordau, Max, 41
Norman-Neruda, Wilma, 54–56; caricature of, 59, *59*
Novello's Music Primers, 44
Nussbaum, Martha, 219n9

Offences Against the Person Act (1861), 103, 205n76
On the Musically Beautiful (Hanslick), 40
On the Sensations of Tone as a Physiological Basis for the Theory of Music (Helmholtz), 10, 28–29, 34–36, 107
"Orchestra" skit on SNL, 189–90
"On the Origin and Function of Music" (Spencer), 31

Paderewski, Jan Ignace, 70
Paganini, Niccolò, 67, 120
Parke, William T., 59
Parsons, Flora T., 46, *46*
Pater, Walter, 128
Peraino, Judith, 17

perfect pitch, 92–93
Perlman, Itzhak, 194
phonograph, 166
physical acoustics, 2; Chladni experiments in, 27–28, *28*; Helmholtz experiments in, 28–29; Tyndall experiments in, 29–31
physiological acoustics, 2, 7, 10, 27; in circulatory and respiratory systems, 36–37; Curtis experiments in, 32; ear and, 32–36; evolutionary science and, 31; Helmholtz experiments in, 34–36; in music curricula, 44
physiological aesthetics, 209n83; acousticians and, 39–40, 210n91; debates about, 37–42
Physiological Aesthetics (Allen), 39
pianola, 166
Picker, John, 9, 34, 107, 108
The Picture of Dorian Gray (Wilde, O.), 13, 129
Pierre (Melville), 148
Pitt, Percy, 60
pornography: music and, 129–133; *Teleny* as, 128, 228n3; in Victorian England, 129, 229n15
posthumanism, 150, 164, 233n18
Prime-Stevenson, Edward, 129
Prosser, Jay, 12
Przybylo, Ela, 3, 16, 76, 83

Quasthoff, Thomas, 191
queer kinship, 6–8, 151, 157
queer musicology, 7, 14–15, 17, 82–83, 141; castrati in, 171, 237n13; human-instrument relations in, 152–153
queer sadness, 2, 20, 173, 180–81
queer temporality, 20, 172–73
queer theory, 2, 3–5; antinormativity and, 16; erotic and, 3–4, 15–17, 200n15; ghosts in, 172–73; posthumanism in, 150, 233n18
queer use, 5, 11–12, 17, 134

race and classical music, 7–8, 193–98

Radau, Rodolphe, 37
Radcliffe, Ann, 148
rape, 6, 8, 13; feminist theories of, 103–5, 226n74, 2234n23; rape culture and, 106, 108–9; in Victorian England, 103, 122. *See also* musical rape
Rebelo, Pedro, 151, 153
Relaxed Performance Project, 195–96
reparative readings, 2
repetitive strain injuries (RSIs), 194
resonator, 34, 35
Richardson, Samuel, 104
Riddell, Fraser, 12, 15, 129, 130, 148
Robert Elsmere (Ward), 18, 51–53; acoustical science and, 52, 62–69; marriage in, 69, 217n97; physiological acoustics and feminist politics in, 77–78
The Romance of the Forest (Radcliffe), 148
Romanes, George, 116
Romantic (music philosophy), 17, 37–38, 208n71
Romola (Eliot), 101
Ross, Alex, 38, 190
Royal Dispensary for Diseases of the Ear, 32
RSIs. *See* repetitive strain injuries
Ruskin, John, 40
Ryan, Vanessa, 109, 225n59

sadomasochistic (S/M) practices, 143
Saint-George, Henry, 67
same-sex desire. *See* homosexuality
Sanden, Paul, 152
Santos, Carles, 152–53
Saturday Night Live (SNL), 189–90
Scarry, Elaine, 116
Scharwenka, Xaver, 153
Schumann, Robert, 137
Schytte, Trida, 55
Scott, Marion, 56, 148
Seashore, Carl, 64
Sedgman, Kirsty, 194, 209n77
Sedgwick, Eve Kosofsky, 2, 4, 8

Seiler, Emma, 29
Senf, Carol, 90
sexology (music and), 12, 15, 37, 129
sexual selection (music and), 31, 129
sexual violence. *See* musical rape; rape
Shaw, George Bernard, 166
Shinner, Emily, 55
Sidgwick, Cecily, 54
Simpson, Hannah, 195–96
"Sinfonia Eroica" (Levy), 1–5, 13, 199n7; erotic in, 3–4
The Sins of the Cities of the Plain (anonymous), 130
siren, 53–54
S/M. *See* sadomasochistic practices
Smith, W. Macdonald, 154
Smyth, Ethel, 128
SNL. See *Saturday Night Live*
Society of Women Musicians (SWM), 60, 98
Somerville, Mary, 38
sonic rape. *See* musical rape
sonic torture. *See* music torture
"Sonorous Sensitive!!!!," 26
sound, 2; figure of, 12–13; music *vs.*, 2
Sound (Tyndall), 29
Sound and Music (Taylor), 44, 212n127
sound studies, 9–10
sound waves, 2, 9, 25–27, 35, 44–45, 64–65, 108, 182
Spampinato, Erin, 19, 103, 105, 114, 122
Spencer, Herbert, 12, 73, 115–16, 129, 162; "On the Origin and Function of Music" by, 31
Spohr, Louis, 63
Steege, Benjamin, 3, 43; on Helmholtz, 34, 36, 208n48; on music education, 46
Sterne, Jonathan, 32
Stone, W. H., 44, 155
Stott, Andrew, 170
Stradivari, Antonio, 157
"Stradivarius" (Eliot), 157

Sully, James, 39–40, 210n86
Sutcliffe, Wallace, 60, 80
SWM. *See* Society of Women Musicians
Symonds, John Addington, 128
sympathetic kinship, in *The Mill on the Floss*, 157–60
sympathetic vibration: in "The Fiddler of the Reels," 121; in *The Lost Stradivarius*, 176, 239n46; in *On the Sensations of Tone* (Helmholtz), 29, 34–35; in *Robert Elsmere*, 65; in *Teleny*, 134–35, 138

Taylor, Clare, 220n16
Taylor, Jodi, 7, 17, 83, 104–5, 133, 141. *See also* aural sex
Taylor, Sedley, 44, 212n127
Tchaikovsky, Pyotr Ilyich, 128
technicon, *154*, 154–155
The Technics of Violoncello Playing (van der Straeten), 57
Teleny (anonymous), 6, 12, 16, 19; aural sex in, 133, 141–43, 231n69; defense of same-sex desire in, 133–41; earworm in, 136–38, 140; exoticization in, 135–46; as gay porn novel, 128, 228n3; Hirsch translation of, 140, 142; orgasm in, 127–28, 131–32, 135; as "A Physiological Romance," 131; publication history of, 128–31, 228n28, 229n15; sympathetic vibration in, 134–35, 138
Tennyson, Alfred, 9
"The Tenor and the Boy" (Grand), 84
Tess of the D'Urbervilles (Hardy), 104, 117, 122, 223n18
Thomas, Bertha, 5–6, 91–92. *See also The Violin-Player*
Thomas, Frances, 91, 221n50
Thomas, Theodore, 191
Tissot, James, 55
Tod, David, 96
Tommasini, Anthony, 193
Toner, John Baptist, 93

tones: compound or partial, 177–78; pure, 93–94
Tonic Sol-Fa exercises, 46, 213n141
Toop, David, 169–70
Touretteshero, 195
trans musicology, 3, 12, 16, 82–83
Trollope, Anthony, 160, 167. *See also The Warden*
Trower, Shelley, 31, 207n31
Tua, Teresina, 55
Tyndall, John, 2, 9; on Chladni, 27; Christmas Lecture by, 25, *26*, 27, 65; glass ball demonstration by, 29–31, *30*; Helmholtz and, 29; on music *vs.* noise, 29; on partial tones, 177–78; physiological studies by, 32, 35; *Sound* by, 29

Under the Greenwood Tree (Hardy), 163
Upton, George, 96–97
Urso, Camilla, 61–62, 96–97

Vaillant, Gabrielle, 91
van der Straeten, Edmund S. J., 57, 153
Venus and Tannhäuser (Beardsley), 130
Vicinus, Martha, 85
"Viola da Gamba" (von Bayros), 130, *131*
"The Violin" (Marsh), 169
The Violin Lady (Campbell, D.), 54, 56
Violin Verses (Scott), 148
The Violin-Player (Thomas, B.), 18; gender politics in, 79–84, 91–98; marriage in, 97, 222n87; violin as companion in, 97, 148, 222n91

Wagner, Richard, 38, 40, 190, 192; Eliot and, 107; Hardy and, 115; Lee and, 181; in *Robert Elsmere* (Ward), 63, 64
Wakefield, Mary, 181
Wang, Yuja, 193, 197
Ward, Mary Augusta, 16; *The Case of Richard Meynell* by, 69; musical

Ward, Mary Augusta *(continued)* interest of, 62. See also *Robert Elsmere*

The Warden (Trollope), 20; human-instrument encounters in, 149, 160–63; muscle memory in, 162–63

Weliver, Phyllis: on the English Musical Renaissance, 199n8; on *The Mill on the Floss*, 108, 114, 159; on music and associationist psychology, 211n101, 218n119; on music-literature studies, 203n50; on *Robert Elsmere*, 62, 215n57, 217n97; on "Sinfonia Eroica," 199n7; on women musicians, 15, 53

Western classical music, 7, 12, 201n26. *See also* classical music (21st century)

Wheatstone, Charles, 28

"A Wicked Voice" (Lee), 20; acoustical science in, 170–73; physiological acoustics in, 180–87; queer suffering in, 180–81; queer temporality in, 20, 172–73, 186

Wilbourne, Emily, 237n13

Wilde, Oscar, 228n3; cello jacket of, 128; imprisonment of, 13, 134; *The Picture of Dorian Gray* by, 13, 129

Wilde, William Robert, 32

Willey, Angela, 12

Wilson, Elizabeth, 77

Winn, Edith Lynwood, 55, 56

"Winthrop's Adventure" (Lee), 240n90

"The Witch of Atlas" (Shelley), 220n31

Wolfreys, Julian, 170, 171

Wollaston, William Hyde, 96

Wonders of Acoustics (Radau), 37

Wong, Amy R., 7

Wood, Elizabeth, 129

Wood, Henry, 56, 60

Zevin, Gabrielle, 147–48

Zon, Bennett, 46